OCEANS

SEALIFE
EXPLORATION
WORLD GEOGRAPHY

Amanda Bennett

This book belongs to

_____

Oceans:
Sealife, Exploration, World Geography

ISBN 1-888306-08-4

Copyright © 1996 by Amanda Bennett

Published by:
Homeschool Press
229 S. Bridge St.
P.O. Box 254
Elkton, MD 21922-0254

Send requests for information to the above address.

Cover design by Mark Dinsmore.

animation by Jeff Bruette

Printed in the United States of America.

This book is dedicated to
beach walkers,
sand castle builders,
and children who ask endless questions!

How To Use This Guide

Welcome to the world of unit studies! They present a wonderful method of learning for all ages and it is a great pleasure to share this unit study with you. This guide has been developed and written to provide a basic framework for the study, along with plenty of ideas and resources to help round out the learning adventure. All the research is done. These are READY to go!

TO BEGIN: The <u>Outline</u> is the study "skeleton", providing an overall view of the subject and important subtopics. It can be your starting point— read through it and familiarize yourself with the content. It is great for charting your course over the next few weeks (or developing lesson plans). Please understand that you do not necessarily have to proceed through the outline in order. I personally focus on the areas that our children are interested in first—giving them "ownership" of the study. By beginning with their interest areas, it gives us the opportunity to further develop these interests while stretching into other areas of the outline as they increase their topic knowledge.

By working on a unit study for five or six weeks at a time, you can catch the children's attention and hold it for valuable learning. I try to wrap up each unit study in five or six weeks, whether or not we have "completed" the unit outline. The areas of the outline that we did not yet cover may be covered the next time we delve into the unit study topic (in a few months or perhaps next year). These guides are <u>non-consumable</u>—you can use them over and over again, covering new areas of interest as you review the previous things learned in the process.

The <u>Reading and Reference Lists</u> are lists of resources that feed right into various areas of the <u>Outline</u>. The books are listed with grade level recommendations and all the information that you need to locate them in the library or from your favorite book retailer. You can also order them through the national Inter-Library Loan System (I.L.L.)—check with the reference librarian at your local library.

There are several other components that also support the unit study.

The <u>Spelling and Vocabulary Lists</u> identify words that apply directly to the unit study, and are broken down into both Upper and Lower Levels for use with several ages.

The <u>Suggested Software, Games and Videos Lists</u> includes games, software and videos that make the learning fun, while reinforcing some of the basic concepts studied.

The **Activities and Field Trip Lists** include specific activity materials and field trip ideas that can be used with this unit to give some hands-on learning experience.

The **Internet Resources List** identifies sites that you might find helpful with this unit. The Internet is a wonderful resource to use with unit studies providing the sights and sounds of things that you might never otherwise experience! You can see works of art in the Louvre. See the sunrise on Mt. Rushmore, hear the sounds of the seashore and find many other things that will help provide an "immersion" in the unit study topic, as never before, without ever leaving home. As with any resource, use it with care and be there with the students as they go exploring new learning opportunities.

The author and the publisher care about you and your family. While not all of the materials recommended in this guide are written from a Christian perspective, they all have great educational value. Please use caution when using any materials. It's important to take the time to review books, games, and Internet sites before your children use them to make sure they meet your family's expectations.

As you can see, all of these sections have been included to help you build your unit study into a fun and fruitful learning adventure. Unit studies provide an excellent learning tool and give the students lifelong memories about the topic and the study.

Lots of phone numbers and addresses have been included to assist you in locating specific books and resources. To the best of our knowledge, all of these numbers were correct at the time of printing.

The left-hand pages of this book have been left "almost" blank for your notes, resources, ideas, children's artwork, or diagrams from this study or for ideas that you might like to pursue the next time you venture into this unit.

"Have fun & Enjoy the Adventure!"

Table of Contents

Introduction

Remember the first time that you "heard" the ocean in a seashell, or felt the sand shift from beneath your feet as the surf washed by? The cry of the gull, the steady rhythm of the surf on the shore, the sting of salt on your face—all of these and more remind us of the tranquility you can experience when seeing the ocean. While the ocean can appear very peaceful to the casual observer, it is teeming with life and has a colorful history that we can not afford to overlook. From the sight of a breaching whale, to the joyful rolling of the dolphin in the waves, to the sounds of the creaking and groaning of a wooden ship under full sail on its way to Treasure Island, you and your family will enjoy studying the vast oceans of the world.

Because the oceans cover more than two-thirds of this planet and have been around since creation, they provide a broad study topic for all ages. When you begin the study, select a few subtopics that are of interest to your student(s) and plan on revisiting this study topic several times over the next few years to learn more about the many facets of the ocean. This study includes:

- History and exploration
- World geography
- Oceanography
- Marine biology
- Ships, lighthouses and more!

Many of you have been able to visit the ocean, while some of you or your children have not yet had that opportunity. It is my hope that this study brings you closer to "seeing" the ocean from your home and getting a feel for its amazing beauty and importance. A study of the ocean will provide a lasting awareness of the delicate balance between all of the elements of creation; animal life, plant life, the food chain, weather, and water.

Because we are fortunate enough to live close to the ocean, I have tried to share what we see and learn through this book—to put together the learning resources that will enable you to share the ocean with your own children and provide a lifetime of memories and knowledge. Grab a towel, a pail and shovel, throw some sand in the back yard and come join the quest. As always, enjoy the adventure!

Study Outline

I. Introduction

    A.   Definition and origin of the words "ocean" and "sea"

    B.   Importance of studying the ocean

        1.   Vast resources for the future

        2.   Keys to our future as well as the past

        3.   Understanding the world and the delicate balance that God created

II. History and Exploration

    A.   Creation

        1.   "In the beginning God created the heaven and the earth." Genesis 1:1 (KJV)

        2.   "And God called the dry land Earth; and the gathering together of the waters called he Seas: and God saw that it was good." Genesis 1:10 (KJV)

        3.   "And God said, Let the waters bring forth abundantly the moving creature that hath life..." Genesis 1:20 (KJV)

        4.   "And God created great whales, and every living creature that moveth, which the waters brought forth abundantly, after their kind, and every winged fowl after his kind: and God saw that it was good." Genesis 1:21 (KJV)

        5.   "For in six days the Lord made heaven and earth, the sea, and all that in them is..." Exodus 20:11 (KJV)

    B.   Exploration

        1.   Early travels

            a.   Primitive boats

            b.   Early routes of water travel

        2.   Early explorers

            a.   Phoenicians

            b.   Greeks

            c.   Romans

            d.   Norsemen (Vikings)

        3.   Explorers in the Age of Discovery

            a.   Henry the Navigator (Prince Henry) of Portugal

            b.   Christopher Columbus

            c.   Vasco da Gama

    d. Bartholomew Diaz
    e. John Cabot
    f. Vasco Nuñez de Balboa
    g. Juan Ponce de León
    h. Hernando Cortez
    i. Ferdinand Magellan
    j. Francisco Vasquez de Coronado
    k. Hernando de Soto
    l. Francis Drake
    m. Walter Raleigh
   4. Later ocean explorers
    a. Abel Tasman
    b. James Cook
    c. Jacques Piccard
    d. Donald Walsh
    e. Jacques-Yves Cousteau

III. Geography

 A. Earth
   1. Dimensions
    a. Circumference
    b. Diameter
    c. Distance from the sun
    d. Volume
    e. Mass
   2. More than 70% of the Earth's surface is covered by water
   3. Magnetic poles
 B. Oceans
   1. Major oceans
    a. Atlantic Ocean
    b. Pacific Ocean
    c. Indian Ocean
    d. Arctic Ocean
    e. Antarctic Ocean
   2. Continental shelf
    a. Components
    b. Geographical features

3. Geographical features of the ocean floor
    a. Plates
    b. Mountain ranges
    c. Abyssal basins
    d. Abyssal plains
    e. Seamounts
    f. Trenches
    g. Fracture zones
4. Islands
    a. Atolls
    b. Volcanic activity
    c. Major island groups
5. Other geographical features
    a. Coral reefs
    b. Deep sea vents
C. Continents
    1. North America
    2. South America
    3. Europe
    4. Asia
    5. Australia
    6. Antarctica
    7. Africa
D. Maps
    1. Uses
        a. Direction (North, South, East, West)
        b. Location
            (1) Lines of latitude
                (a) Horizontal
                (b) Includes the equator
            (2) Lines of longitude
                (a) Vertical - from North to South Poles
                (b) Includes the Greenwich Meridian
        c. Surface characteristics
            (1) Topography
            (2) Geographical characteristics
            (3) Other features of an area (for example: population, rainfall amounts, ocean currents, etc.)

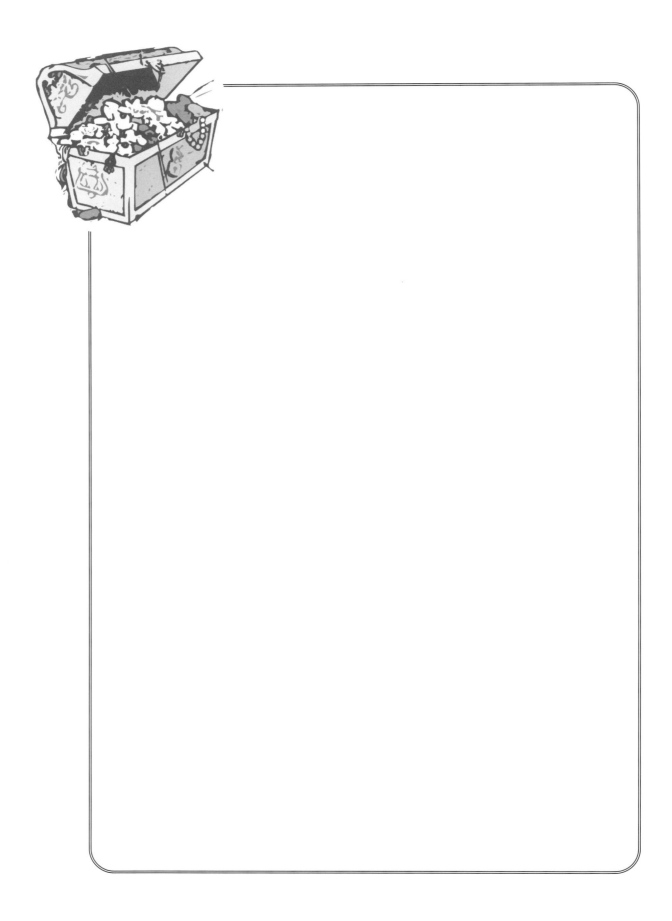

2. Features of a map
    a. Scale
    b. Legend
    c. Indicator of the direction of North

IV. Oceanography

A. Movement of the water
    1. Currents
        a. Types of currents
            (1) Surface currents
            (2) Subsurface currents
        b. Causes of currents
            (1) Coriolis force
            (2) Wind on the surface
            (3) Water density differences
    2. Tides
        a. Tidal cycle
            (1) High tide
            (2) Low tide
        b. Astronomical tides
            (1) Neap tide
                (a) Lower-than-normal high tide
                (b) Occur when sun and moon are at right angles
            (2) Spring tide
                (a) Higher-than-normal high tide
                (b) Occur when sun and moon are lined up with the Earth in a straight line
    3. Waves
        a. Causes of waves
            (1) Wind acting on the surface of the ocean
            (2) Sudden geologic changes under the surface of the ocean—volcanic eruption, earthquake, etc.
        b. Wave description terms
            (1) Crest (top)
            (2) Trough (bottom)
            (3) Height (distance from crest to trough)
            (4) Wavelength (distance between waves)

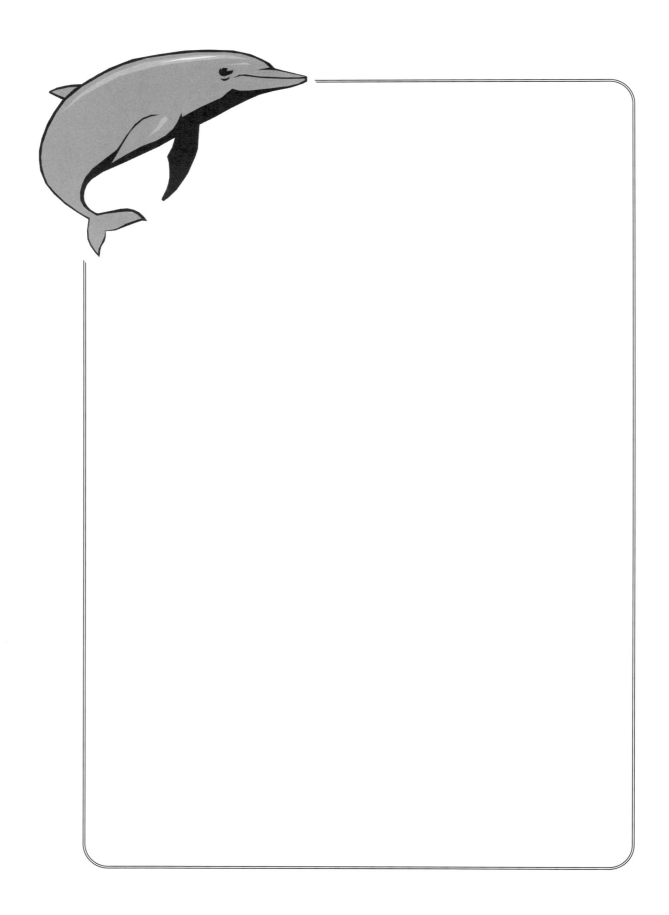

B.   Chemistry of the ocean water
    1.   Chemical components
        a.   Water (hydrogen and oxygen)
        b.   Sodium
        c.   Magnesium
        d.   Calcium
        e.   Potassium
        f.   Sulfates
        g.   Chlorides
    2.   Salinity of the ocean water

C.   Products from the ocean
    1.   Food
        a.   Fish
        b.   Plants (seaweed, etc)
    2.   Petroleum products
    3.   Minerals (manganese nodules, for example)

V.   Ships

A.   History
    1.   Reasons for building ships
        a.   To explore
        b.   To develop trade
        c.   To find other people
    2.   Shipbuilders in history
        a.   Phoenicians
        b.   Vikings
        c.   Greeks
        d.   Romans
        e.   Chinese

B.   Types
    1.   Raft
    2.   Sail-propelled
    3.   Human-propelled
    4.   Engine-propelled
    5.   Remote-controlled submersible

C.   Study of maritime history
    1.   Meaning of maritime
    2.   Impact of ships on history

D. Famous shipwrecks
   1. *Nuestra Señora de Atocha*
   2. *Mary Rose*
   3. *Titanic*
   4. *Lusitania*
   5. *Bismark*
   6. *S. S. Central America*

VI. Navigation

A. Latin origin of the word "navigate"
B. Four basic methods of marine navigation
   1. Celestial navigation—using the sun, moon and stars
   2. Piloting navigation—using visual landmarks and depth soundings
   3. Dead reckoning navigation—using a carefully-maintained record (or "reckoning") of the ship's motion (time, direction and distance)
   4. Electronic navigation—using radar, radio direction finder, etc.
C. Tools of navigation
   1. Early navigational tools
      a. Astrolabe
      b. Sextant
   2. Compass
   3. Nautical charts
   4. Dividers
   5. Navigational lights, markers and buoys

VII. Marine biology (sea life)

A. Three "zones" of life in the ocean
   1. Sunlit zone
   2. Twilight zone
   3. Bathypelagic zone
B. Plant life examples
   1. Phytoplankton
   2. Seaweed

C. Animal life examples
1. Whale
2. Dolphin
3. Sea walrus
4. Seal
5. Tuna
6. Marlin
7. Swordfish
8. Snapper
9. Bonito
10. Shark
11. Sponge
12. Coral
13. Sea turtle
14. Jellyfish
15. Eel
16. Starfish
17. Sea urchin
18. Scallop
19. Sea cucumber
20. Sand dollar
21. Sea anemone
22. Octopus
23. Squid
24. Ray
25. Manatee

VIII. Seashore

A. Beaches
1. Tidal pools
2. Sea shells
3. Sea birds
   a. Sea gull
   b. Pelican
   c. Sandpiper
   d. Heron
4. Land animals that live along the shore
   a. Sand crab (fiddler crab)
   b. Sand flea
5. Shoreline of the beach shifts through various tide changes

B. Recreation
    1. Swimming
    2. Boating
    3. Surfing
    4. Fishing
    5. Scuba diving/snorkeling
    6. Shell collecting
    7. Treasure hunting

IX. Weather and the ocean

A. Hurricanes
B. Tsunamis
C. Typhoons
D. Water spouts
E. El Niño
F. Prevailing winds

X. Lighthouses

A. Purpose
B. Types
C. History
D. Location

XI. Art and ocean

A. Literature
    1. *Moby Dick*
    2. *Swiss Family Robinson*
    3. *Robinson Crusoe*
    4. *Treasure Island*
    5. *Kidnapped*
    6. *20,000 Leagues Under the Sea*
B. Paintings
    1. Seascapes
    2. Ships at sea

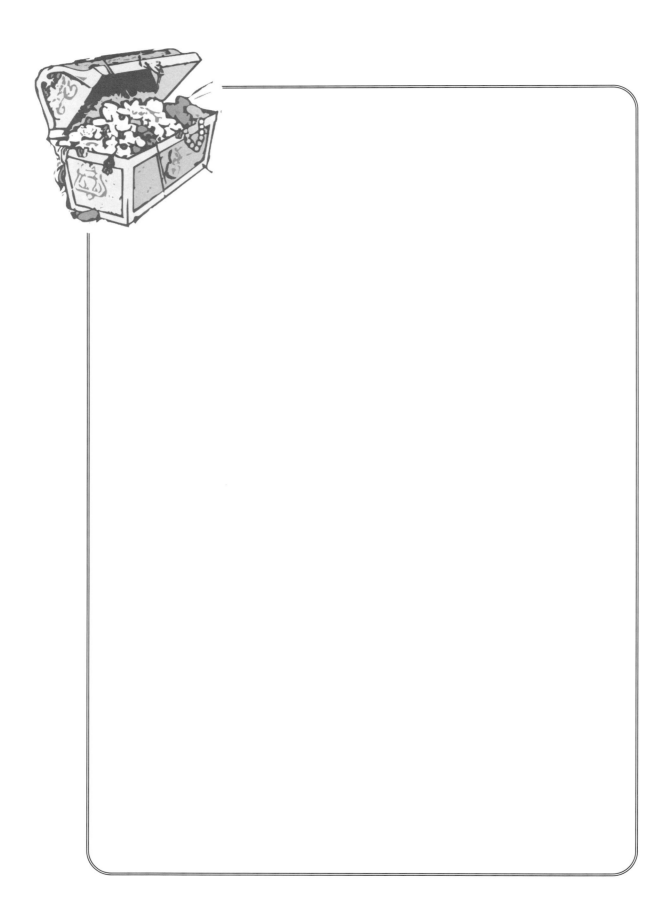

C. Music
1. Sailing songs
2. Sounds of the ocean set to music

XII. Modern ocean research

A. Goals
1. Resources
a. Food
b. Petroleum
c. Minerals
2. Information to improve medical technology
3. Increase our knowledge of the ocean floor and its components
B. Research institutes continue the quest for knowledge of the ocean
1. Scripps Institution of Oceanography
2. Woods Hole Oceanographic Institution
3. Harbor Branch Oceanography Institute

Spelling and Vocabulary
Lower Level

animal	fish
area	fishing
bait	float
bay	flood
beach	floor
bird	food
blow	globe
boat	gulf
body	hook
captain	jellyfish
cast	kelp
chain	knot
clam	krill
cod	life
continent	map
coral	mask
crab	mast
crew	mile
deep	move
dig	net
dive	ocean
diver	pilot
dolphin	plant
drift	pole
eat	rain
eel	ray
egg	reef
explore	river
family	rope
fin	sail

sailor
salt
sand
sea
sea gull
seal
seawater ✓
seaweed ✓
shark

shell
ship
shore
sink
skate
squid ✓
stars
sun
sunken
swim

tank
tide
time
treasure
tuna
turtle
water
wave
whale
wind

Writing Ideas

Here are some ideas to help incorporate writing in a unit study. Choose one or two and watch what happens!

1. While studying some of the ocean explorers, have each student keep an imaginary journal of an exploration trip with one of the famous explorers (Columbus, Balboa, Henry the Navigator, etc.). He can keep track of his daily travel progress, as well as share some of the feelings and sights he might experience during this kind of trip.

2. When reading the **Swiss Family Robinson**, consider having each child dictate or write about the possible adventures he might experience as a castaway on an island. Ask questions like "What would you do first?", "Where would you stay?" and "What would you eat?" Each one can keep a journal of these imaginary adventures, including sketches of the island, his home, etc.

3. Plan a real or imaginary trip to the ocean with your student(s). Have them write to the Chamber of Commerce of your chosen destination. Check with the reference librarian at the local library to obtain the address information. The Chamber might send hotel information, lists of museums and other spots of interest, as well as some historical information about the area. Use these brochures and other collected information to have the children plan an agenda for the trip, as well as a written proposed budget.

4. If the student has been to the seashore, lake or river, have him write or dictate a memory from the trip. He can describe the people, wildlife observed and water fun he had during the visit. If you have any photos, he can include them in his story!

5. If you have ever been to the ocean, consider sharing some of your memories with your student(s)—the cottage you stayed in, the friends that were there, the seashells that you found, the early morning walks in the sunrise, etc. Then, have each of them compare your visit with one that he himself remembers.

6. Consider having your student write or dictate an imaginary journal that is a daily log of his travels on an expedition with a famous explorer like Columbus, Henry the Navigator, etc. He can describe what he might have seen, eaten, experienced and discovered while on his journey.

7. There have been many stories throughout the history of ocean lore concerning messages in bottles that floated along with the ocean currents. A fun project might involve writing a message on a small piece of paper, describing the student's location and circumstances, and hold a mock-bottle-launching. Then, have the student select an island for his adventure. Try to estimate the time and possible landing sites for the bottle, using a map of currents for that particular area of the ocean, as well as a description of who might find their message.

Activity Ideas

Activities are a great way to reinforce the material that we learn. They provide important hands-on learning. This allows the student to have fun and be challenged at the same time. Here are some activity resources and ideas that we found to use with this unit:

1. There are many inexpensive jigsaw puzzles available, at all different levels of difficulty, that have ocean-oriented themes. Some of these include whales, dolphin, ships, seascapes, beach scenes, harbors, fish, etc. These are fun to work on together as a family, on a rainy day or during a read-aloud session using one of the fun classics like **Treasure Island**.

2. As you study the various kinds of ships and boats used on the ocean, look for small hobby models available in many discount stores. You can often find models of many kinds of ships, including the tall ships, racing boats, submarines, and so on. These can be fun to work on either individually by the children or as a family project.

3. As you study the geography of the ocean, consider having the students build models of an island, volcano, shoreline or the ocean floor.

4. Fishing is a great hobby to use to learn more about life in the oceans, as well as tides and other facets of the ocean. With your students, experiment with either salt or fresh-water fishing, or both, if available. Have the student keep a fishing journal that keeps track of the tides, types of bait, successes and ideas for the next fishing trip.

5. Using clay, craft sticks or various building blocks, have the children construct a model of a lighthouse. They can use one from a picture as a guide, or use their imagination to create one from **Keep the Lights Burning Abbie** or the Boxcar Children's **Lighthouse Mystery**.

6. If possible, have the children develop a collection of ocean-related items. These might include sea shells, post cards, rocks, driftwood, etc. If they collect rocks or shells, have them identify each specimen and label each one with the location and date of the find.

7. When reading **The Swiss Family Robinson**, have your children draw and then build a model of the one of the homes described in the story.

Activity Resources

EXPERIMENTS:

Sink or Swim: The Science of Water, Simple Experiments for Beginners!, by Barbara Taylor. (Step Into Science Series). Grades 2–5. Published by Random House Books for Young Readers, 400 Hahn Rd., Westminster, MD 21157. (800) 733-3000.

Rivers & Oceans: Geography Facts & Experiments, by Barbara Taylor. (Young Discoverers Series). Grades 1–4. Published by Larousse, Kingfisher, Chambers, Inc. 95 Madison Avenue, New York, NY 10016. (800) 497-1657.

The Ocean Book: Aquarium and Seaside Activities for all Ages, by the Center for Marine Conservation staff. (Grades PreK–6). Published by John Wiley & Sons, 1 Wiley Dr., Somerset, NJ 08875. (800) 225-5945, ext. 2497.

Milliken Science Transparency Reproducible Books, published by Milliken Publishing Company. Grades 4–9. Available from Farm Country General Store, Rt. 1, Box 63, Metamora, IL 61548. (800) 551-FARM. Titles to consider with this unit include:
Oceanography and **Fish, Amphibians & Reptiles**

How to be an Ocean Scientist in Your Own Home, by Seymour Simon. Grades 5 - 9. Published by HarperCollins Children's Books, 1000 Keystone Industrial Park, Scranton, PA 18512. (800) 242-7737.

Janice VanCleave's Geography for Every Kid, Janice VanCleave's Earth Science for Every Kid, and **Janice VanCleave's Oceans for Every Kid,** all by Janice Van Cleave. Grades 4 and up. Published by John Wiley & Sons, 1 Wiley Dr., Somerset, NJ 08875. (800) 225-5945, ext. 2497.

Projects in Oceanography, by S. Simon (Science at Work Series) Grades 5–8. Published by Franklin Watts, Chicago, IL (800) 672-6672

HANDS-ON ACTIVITIES:

Easy-To-Make Lighthouse, by Edmund V. Gillon. Grades 5 and up (adult assistance required for the younger students). Published by Dover Publications, 31 East 2nd Street, Mineola, NY 11501.

Easy-To-Make Noah's Ark, by A.G. Smith. Grades 3 and up (adult assistance required for the younger students). Published by Dover Publications, 31 East 2nd Street, Mineola, NY 11501.

How To Make a Clipper Ship Model, by Armitage McCann. Grades 9 and up. Published by Dover Publications, 31 East 2nd Street, Mineola, NY 11501.

Ship Models: How To Build Them, by Charles Davis. Grades 9 and up. Published by Dover Publications, 31 East 2nd Street, Mineola, NY 11501.

Under The Sea Activity Book, by Jennifer Riggs. Grades PreK–3. Published by Scholastic, Inc., P.O. Box 7502, Jefferson City, MO 65102. (800) 325-6149.

Create Your Own.. Sticker Picture Series, from Dover Publications. Grades PreK–3. Titles to consider for this unit study include:

> **Create Your Own Noah's Ark Sticker Picture**, by Jill Dubin.
> **Create Your Own Beach Sticker Picture**, by Robbie Stillerman.
> **Create Your Own Undersea Adventure Sticker Picture**, by Steven J. Petruccio.
> **Create Your Own Pirate Adventure Sticker Picture**, by Steven J. Petruccio.

Cut and Assemble An Early American Seaport, by A.G. Smith. Grades 8–12. Published by Dover Publications, 31 East 2nd Street, Mineola, NY 11501.

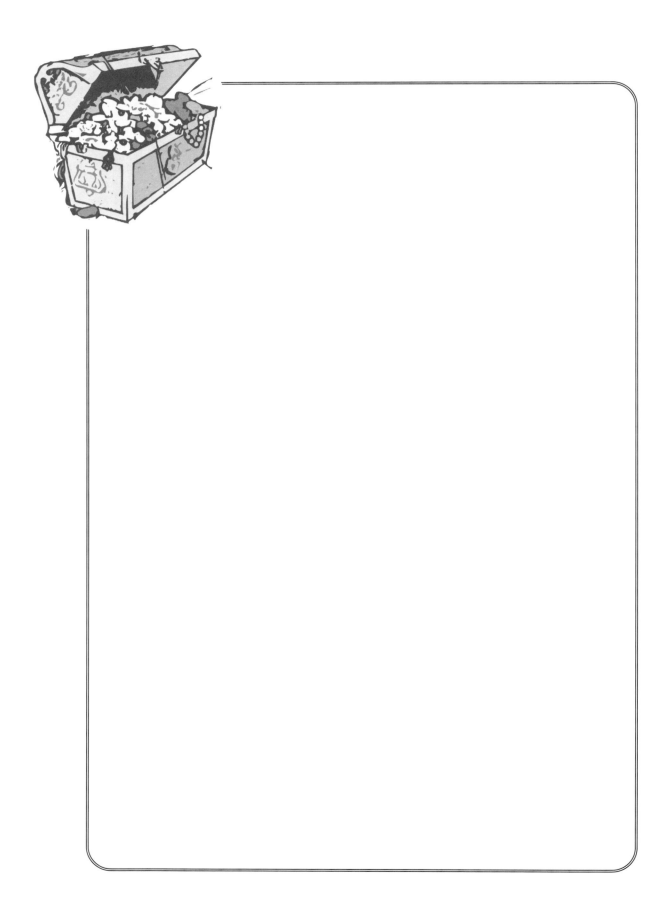

COLORING/DRAWING BOOKS:

The Marine Biology Coloring Book, by Thomas Nielsen. Grades 7–12. Published by HarperCollins, 1000 Keystone Industrial Park, Scranton, PA 18512. (800) 242-7737. Available from Farm Country General Store, Rt. 1, Box 63, Metamora, IL 61548. (800) 551-FARM.

The Peterson Field Guide Coloring Books, published by Houghton Mifflin Company. Grades 3–9. Available from The Elijah Company, Route 2 Box 100-B, Crossville, TN 38555. (615) 456-6384. Titles to consider for this unit study include: **Seashores, Fishes,** and **Shells**

Coloring Book Series from Dover Publications. Grades 3–12. This company publishes a broad range of high-quality, detailed coloring books that are inexpensive and educational for ALL ages. Write to them and request a free catalog at Dover Publications, 31 East 2nd Street, Mineola, NY 11501. Titles to consider for this unit study include:

Marine Biology:
Seashore Life, by Anthony D'Attilio.
Coral Reef, by Ruth Soffer.
Shells of the World, by Lucia de Leiris.
Whales and Dolphin, by John Green.
Sharks of the World, by Lyn Hunter.
Tropical Fish, by Stefen Bernath.
Fishes of the North Atlantic, by Thomas C. Quirk, Jr.

History:
The Story of Whaling, by Peter F. Copeland.
North American Lighthouse, by John Batchelor.
American Sailing Ships, by Peter F. Copeland.
Shipwrecks and Sunken Treasures, by Peter F. Copeland.
Historical Sailing Ships, by Tre Tryckare Co.
Story of the Vikings, by A.G. Smith

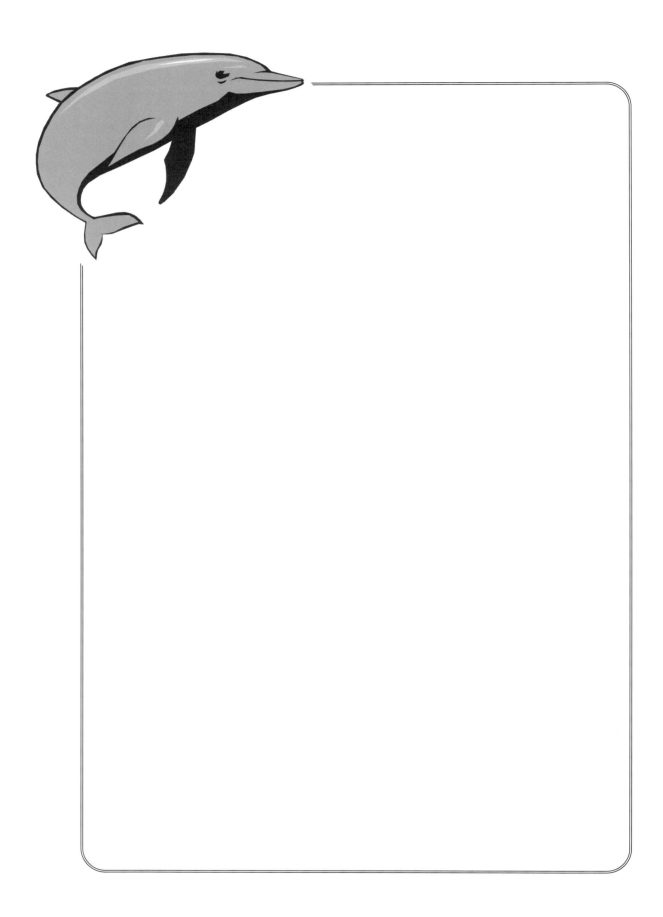

Educational Read and Color Book Series from Spizzirri Publishing, Inc. A fun, inexpensive and educational series of books with optional cassette tapes, available from Spizzirri Publishing, P.O. Box 9397, Rapid City, SD 57709. (800) 322-9819. Titles to consider for use with this unit include:

- **Dolphins** ✓
- **Atlantic Fish**
- **Pacific Fish**
- **Fish**
- **Deep-Sea Fish**
- **Marine Mammals**
- **Penguins** ✓
- **Sharks**
- **Ships**
- **Whales**

Internet Resources

Here are some interesting sights (grouped by study topic) on the Internet that you might want to visit while studying this unit. Please keep in mind that these pages, like all web pages, change from time to time. I recommend that you visit each sight first, before the children do, to view the content and make sure that it meets with your expectations. Also, use the **Subject Key Words** for search topics on Internet search engines, to find the latest additions that might pertain to this topic.

Creation:

Creation Science Homepage
http://emporium.turnpike.net/C/cs

The Creation Research Society
http://www.iclnet.org/pub/resources/text/crs/crs-home.html

Creation Research Foundation
http://users.aol.com/CRFCRESCI/crf.html

Creation Science Association of Atlantic Canada
http://www.navnet.net/csaac/csaac.html

Center for Scientific Creation Home Page
http://www.indirect.com/www/wbrown/

Creation Station Home Page
http://schdist23.bc.ca/MarkMcK/creation.html

Answers in Genesis
http://www.christiananswers.net/aig/aighome.html

The Creation Index
http://www.christiananswers.net/menu-ac1.html

Oceanography:

Links to Oceanographic Information
http://oraac.gsfc.nasa.gov/%7Erienecke/po_others.html

Oceanographic Collections—Scripps Institution
http://gdcmp1.ucsd.edu/sci_coll.html

Woods Hole Oceanographic Institution
http://www.whoi.edu/

Scripps Institution of Oceanography
http://www.sio.ucsd.edu/

Scripps Institution of Oceanography
Oceanography on the Net
http://www.scilib.ucsd.edu/sio/

Sea Surface Temperature Satellite Images
http://dcz.gso.uri.edu/avhrr-archive/archive.html

WWW Tide/Current Predictor - Site Selection
http://tbone.biol.sc.edu/tide/sitesel.html

Other Oceanographic Information Sources
http://www.cms.udel.edu/other/other.html

Welcome to Harbor Branch Oceanographic Institution
http://www.hboi.edu/

NOAA Coastal Estuarine Oceanography Branch
http://www-ceob.nos.noaa.gov/

Interactive Marine Forecast
http://thunder.met.fsu.edu/~nws/buoy/

Marine Biology:

Division of Fishes - Museum of Zoology Web Server
http://ummz1.ummz.lsa.umich.edu/fishdiv/

Alaska Department of Fish and Game Home Page
http://www.state.ak.us/local/akpages/FISH.GAME/adfghome.htm

Northwest Fisheries Science Center
http://research.nwfsc.noaa.gov/

U.S. Fish and Wildlife Service Home Page
http://www.fws.gov/

Australia's On-line Fish File
http://www.ph.unimelb.edu.au/%7Edjd/fishing/fish.html

Smithsonian Museum of Natural History - Division of Fishes
http://nmnhwww.si.edu/vert/fish.html

Sea World Home Page
http://www.bev.net/education/SeaWorld/

Sea Urchins
http://seaurchin.org/Sea-Grant-Urchins.html

Whales on the Net
http://whales.magna.com.au/home.html

Boston Harbor Whale Watch
http://www.bostonwhale.com/

Hyannis Whale Watcher
http://www.capecod.net/Whales/docs/whale2.html

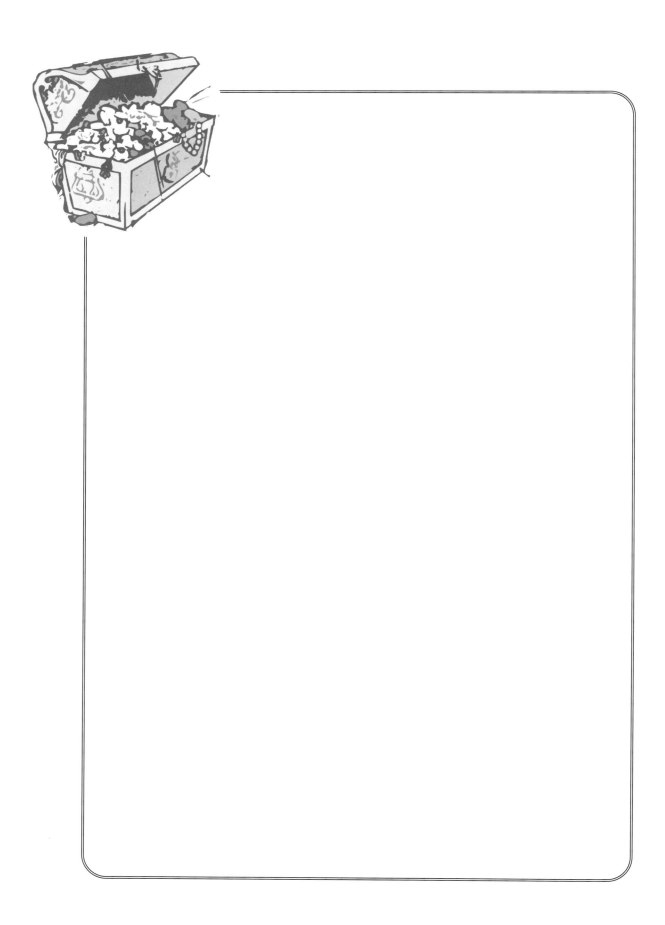

Aquarium/History/Museum Collections:

The Maritime History Virtual Archives
http://pc-78-120.udac.se:8001/WWW/Nautica/Nautica.html

Maritime History on the Internet
http://ils.unc.edu/maritime/home.html

Ferdinand Magellan
http://www.nortel.com/english/magellan/ferdinand/MagellanBio.html

Smithsonian: Ocean Planet Homepage
http://seawifs.gsfc.nasa.gov/ocean_planet.html

The Florida Aquarium
http://www.sptimes.com/aquarium/FA4.htm

New England Aquarium
http://www.neaq.org

Monterey Bay Aquarium: Exhibit Information
http://www.usw.nps.navy.mil/~millercw/aq/exhibit.html

The Intrepid Sea-Air-Space-Museum Home Page
http://www.westnet.com/intrepid/

Naval Undersea Museum
http://www.tscnet.com/tour/museum/index.html

Peabody Essex Museum: Marine Paintings and Drawings
http://www.pem.org/maritm3.htm

NCDA - Maritime Museum
http://www.agr.state.nc.us/maritime/index.htm

Ships, Boats, Submarines and Other Related Items:

Permanent Ship Collection at the Museum
http://www.westnet.com/intrepid/shipcoll.html

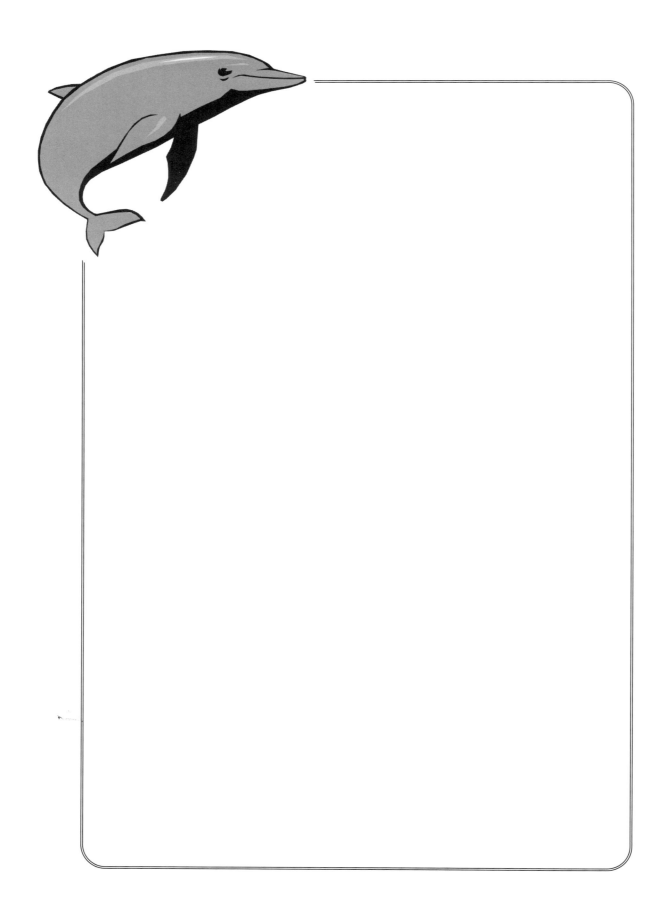

Historic Vessels
http://ils.unc.edu/maritime/ships.html

The Royal Institute Of Navigation Home Page
http://hydrography.ims.plym.ac.uk/rin/rinhome.htm

U.S. Coast Guard Navigation Center
http://www.navcen.uscg.mil/

U.S. Coast Guard—Recreational Boating Safety Information
http://www.navcen.uscg.mil/gnab/gnab.htm

Deep Sea Robots
http://www.mysite.com/deep6/rov.htm

International Maritime Education Schools
http://www.gens.no/educate/

The Sailing Site - News and Information on Sailing, Yacht and Sailboat Racing, Cruising, and Maritime Heritage
http://www.gosailing.com/

Lighthouses and Ships within the National Park System
http://www.cr.nps.gov/history/maritime/maripark.html

Lighthouses:

Lighthouses and Ships within the National Park System
http://www.cr.nps.gov/history/maritime/maripark.html

Inventory of Historic Light Stations
http://www.cr.nps.gov/history/maritime/ltinv.html

Sources for Lighthouse Information
http://www.cr.nps.gov/history/maritime/lightinf.html

World-Wide Web Virtual Library: The World's Lighthouses, Lightships Lifesaving Stations
http://www.maine.com/lights/www_vl.htm

The Lighthouse Society of Great Britain
http://www.soton.ac.uk/~kt1/

Bill's Lighthouse Getaway
http://zuma.lib.utk.edu/lights

Seashells:

Internet Resources for Conchologists
http://fly.hiwaay.net/~dwills/shellnet.html

Shell Pictures
http://www.mindspring.com/~bearl/gashell/shellpic.htm

Miscellaneous:

The Jason Project
http://seawifs.gsfc.nasa.gov/JASON/HTML/JASON.html

Careers in Marine Mammal Science
http://www.bev.net/education/SeaWorld/marinescience.html/mshome.html

MIT Sea Grant AUV Lab
http://web.mit.edu/afs/athena/org/s/seagrant/www/auv.htm

Carolina Biological Supply Company
http://www.carosci.com/

Tarpon Springs—Sponge Industry
http://www.tarponsprings.com/SPG.HTM

Job Opportunities

Here is a list of some of the jobs that involve the oceans. There are others that I'm sure you will identify, but these are some of the main ones that we investigated during our unit study.

Biologist

Chemical engineer

Chemist

Civil engineer

Coast Guard crewman

Electrical engineer

Fisherman/lobsterman ✓

Environmental engineer

Fish and wildlife management
   specialist

Fishery biologist ✓

Geologist

Marine biologist ✓

Marine engineer

Marine geologist ✓

Materials engineer

Mechanical engineer

Merchant marine

Meteorologist ✓

Navigator

Oceanographer

Park ranger

Petroleum engineer

Sailor

Ship captain

Veterinarian

Zoologist ✓

For more information about these jobs or others that may be interesting, go to the reference librarian in the public library and ask for publications on careers. Some that we recommend are:

The Encyclopedia of Careers and Vocational Guidance, published by J. G. Ferguson Publishing Company, Chicago.

Occupational Outlook Handbook, published by the U. S. Department of Labor, Bureau of Labor Statistics. It presents detailed information on 250 occupations that employ the vast majority of workers. It describes the nature of work, training and educational requirements, working conditions and earnings potential of each job.

Room Decorations

When working on a unit study, we try to decorate the room with items that relate to our current topic of interest. This allows the students to see the important information on a regular basis, as well as providing a place to view their own work. For oceans, consider some of the following ideas:

1. Hang a world map on the wall, as well as an inexpensive globe. They can usually be found at the local discount store or bookstore, and will greatly increase the child's understanding of the ocean, as well as Earth's geography during this unit study.

2. There is an abundance of inexpensive posters available that display sea life (whale, dolphin, manatee), ships, lighthouses, etc. I try to collect these as I find them and set them aside for our many forays through the oceans topic. They do indeed add to the "flavor" of our studies!

3. Our local library system offers framed prints on loan for six week intervals. They have some beautiful prints of seascapes, ships, etc., that complement an ocean study.

4. Try to develop a wall "ocean collage", using an inexpensive fish net from an import store as the basic wall hanging. Decorate the net with beach/sea items from previous or ongoing ocean visits, using things like sea shells, driftwood, rope pieces and other debris that may have washed ashore.

5. Have the students draw pictures and/or create posters of the things they are learning—the continental shelf, waves, islands, the ocean floor, shipwrecks, etc.

6. Help the students create displays of their ocean collections or any other collection that they are working on. The oceans collection might be of sharks' teeth, shells, stamps, knots, rope, models, etc.

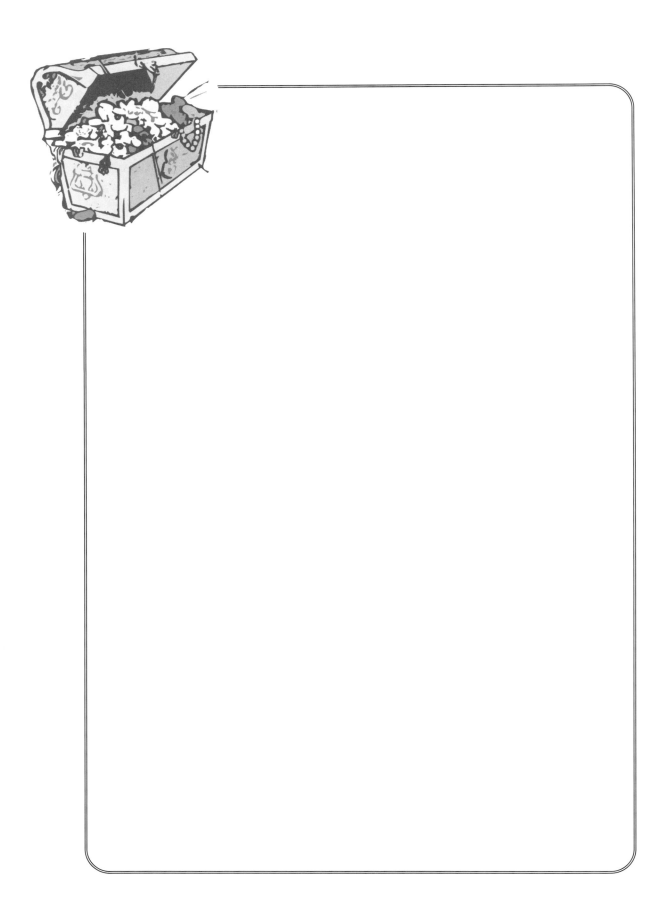

Videos

While your family learns about the oceans, you will find that many great shows and videos are available. Many of the items can be obtained through your local library or video store.

First, check at your local library and see what videos they have in their collection. We found some great ones about sea life, oceanography, ocean exploration, sailing and many others. Some of our favorites were those by National Geographic and various oceanographic institutes.

In addition, there are many great documentaries available on Public Broadcasting Stations and through the library. There are also several episodes of Reading Rainbow that cover oceans and sea life. Many libraries maintain copies of Reading Rainbow episodes, usually in the Juvenile/Young Adult Reference section.

From the Moody Institute of Science:

Roaring Waters, Explore the waters around us. The ebb and flow of tides. Raging rivers. Surging floods. Crashing tidal waves. Marvel at water's role in our infinitely complex weather system and the amazing power of the Master Creator who keeps it all in place. All ages.

Water Water Everywhere, You will learn how water makes the difference between the barren moon and the green earth. One of three movies on a film of science adventures for Grades 1–8.

God's Earth Team, Preserving the Earth He has given you. A delightful way to teach youngsters how to be good stewards of the earth's environment. Grades 1–8.

God of Creation, Explore the power and majesty of the world God created.

> (all are available from Great Christian Books, 229 S. Bridge St., P.O. Box 8000, Elkton, MD 21922-8000, (800) 775-5422)

Really Wild Animals: Deep Sea Dive, by National Geographic Kids Video.

Life in the Sea, by Questar Video. Available from Great Christian Books, 229 S. Bridge St., P.O. Box 8000, Elkton, MD 21922-8000, (800) 775-5422.

Fiction Stories Available on Video:

The Swiss Family Robinson

20,000 Leagues Under the Sea

Treasure Island

Kidnapped

Mutiny on the Bounty

The Old Man and the Sea

Moby Dick

Robinson Crusoe

Games and Software

Games are a great tool to reinforce the material that we learn. We have fun while reviewing important information and concepts around the kitchen table or on the computer. The software listed here is just a small sample of what is available. With the writing of this book, there are several new games and software packages in development for release in the near future, and many sound very exciting! Check around at your local toy and software stores to find out the latest introductions.

Games:
Where in the World, a geography board game from Aristoplay for Grade 3 and up. Published by Aristoplay, P.O. Box 7529, Ann Arbor, Michigan 48107. (800) 634-7738.

GeoSafari, a geography game that is available in either stand-alone game format or on software in CD-ROM format. Grades 3 and up. Produced by Educational Insights, Carson, CA.

Software:
Where in the World is Carmen Sandiego?, a fun detective software game that helps teach world geography. Published by Broderbund Software, 500 Redwood Blvd., Novato, CA 94948.

GeoSafari, a geography game that is available in either stand-alone game format or on software in CD-ROM format. Grades 3 and up. Produced by Educational Insights, Carson, CA.

The Magic Schoolbus Explores the Ocean, an Interactive Science Adventure, available in CD-ROM format. Produced by Microsoft Corporation.

The Nature Collector: Freshwater Fish, a simulation of aquarium management, maintaining tropical fish exhibits, traveling the world to obtain the species you need for your displays. Fun, but challenging! For Grades 3 and up. Available in CD-ROM format. Produced by AnimaTek, Spectrum Holobyte, Attn: Allen, P.O. Box 90848, San Diego, CA 92169. (800) 695-4263.

Field Trip Ideas

There are so many field trips that can be enjoyed while learning about oceans and sea life, that it is hard to list all of the ones that you might want to consider. Please use this list to get started planning some field trips, then let your imagination identify others that may be in your area.

1. First and foremost, if possible, take a day or weekend trip to the ocean shore to let the children take a look around and get a feel for the sights and sounds of the ocean. They can work on a sea shell collection and improve their sandcastle-building skills! If this is not possible, consider taking them to a nearby lake or river for the same kinds of experiences.

2. Another fun and educational field trip is a fishing trip! First, determine the local requirements for fishing licenses for your students. After taking care of licensing matters, take the fishermen to the local bait and tackle shop to get acquainted with the various types of bait and equipment available. Then, set out for a fun day of fishing! Have each student keep notes of the success he has with various types of bait, lures, locations, etc. Hopefully you will all be able to repeat this field trip to various locations to round out your fishing knowledge.

3. Frequently, many of the state fish and game commissions offer free or inexpensive fishing classes for students. At times, they might also offer classes on water safety, boating safety and first aid. Additionally, this unit study presents the perfect opportunity to enroll the children in swimming classes!

4. If there happens to be a boat manufacturer within reasonable driving distance, consider trying to set up a tour of the facilities. Watch boats being assembled and attend tests of new boats and/or motors.

5. Many of the amateur ocean shore treasure hunters use metal detectors while they walk, to locate possible "finds" under the surface. Consider purchasing one from a local pawn shop, an electronics kit assembly catalog, etc—and have your students learn how it works, as well as giving it a good workout in your area.

Subject Word List

This list of **SUBJECT** search words has been included to help you with this unit study. To find material about oceans, go to the card catalog or computerized holdings catalog in your library and look up:

General Words

abyss ✓
Antarctic Ocean ✓
Arctic Ocean ✓
Atlantic Ocean ✓
boating
boats
Coast Guard ✓
compass ✓
continent ✓
coral
crab
creation
dolphin
El Niño ✓
estuary ✓
explorers
fish
fishing
food chain
Gulf Stream ✓
hurricane ✓
iceberg
Indian Ocean ✓
island
jellyfish
kelp ✓
lighthouse
lobster
marine biology ✓
marine science ✓

meteorology ✓
nautical ✓
navigation ✓
ocean travel
ocean
ocean currents
oceanography ✓
Pacific Ocean ✓
pirate
sailing
scuba diving
sea life
sea shells
seaport
shark
ships
shipwreck
shoreline
shrimp
sponge
submarine ✓
submersible ✓
swimming
tidal pool ✓
tides
treasure
typhoon ✓
water
waves
whale

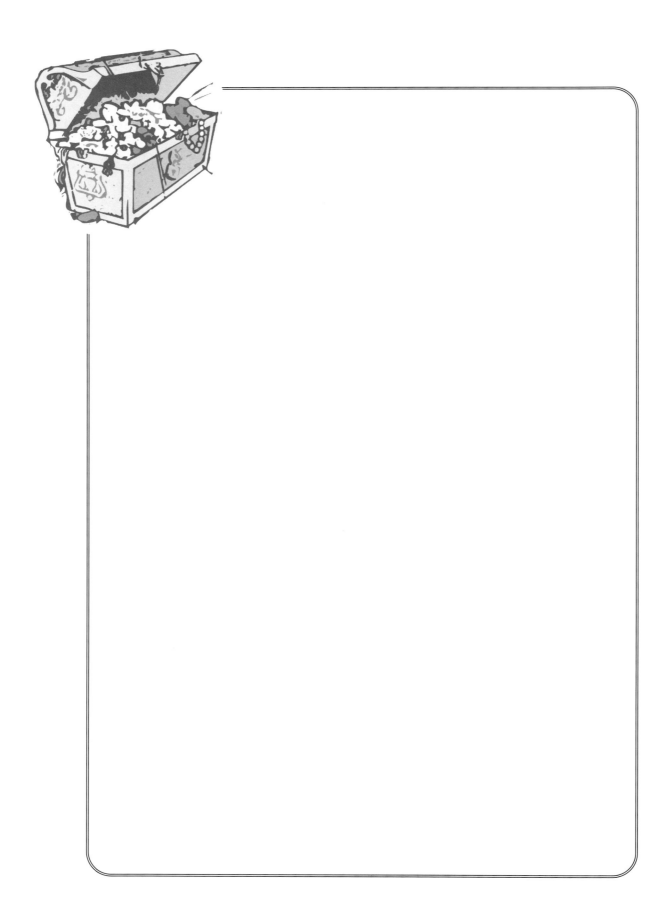

People

Balboa, Vasco Nuñez de

Beebe, Charles William

Bowditch, Nathaniel

Cabot, John

Champlain, Samuel de

Columbus, Christopher

Cook, James

Coronado, Francisco Vasquez de

Cortez, Hernando

Cousteau, Jacques

Da Gama, Vasco

De Soto, Fernando

Diaz, Bartholomew

Drake, Francis

Erickson, Leif

Fulton, Robert

Hudson, Henry

Magellan, Ferdinand

Maury, Matthew F.

Ponce de Leon, Juan

Prince Henry the Navigator

Tasman, Abel

Vespucci, Amerigo

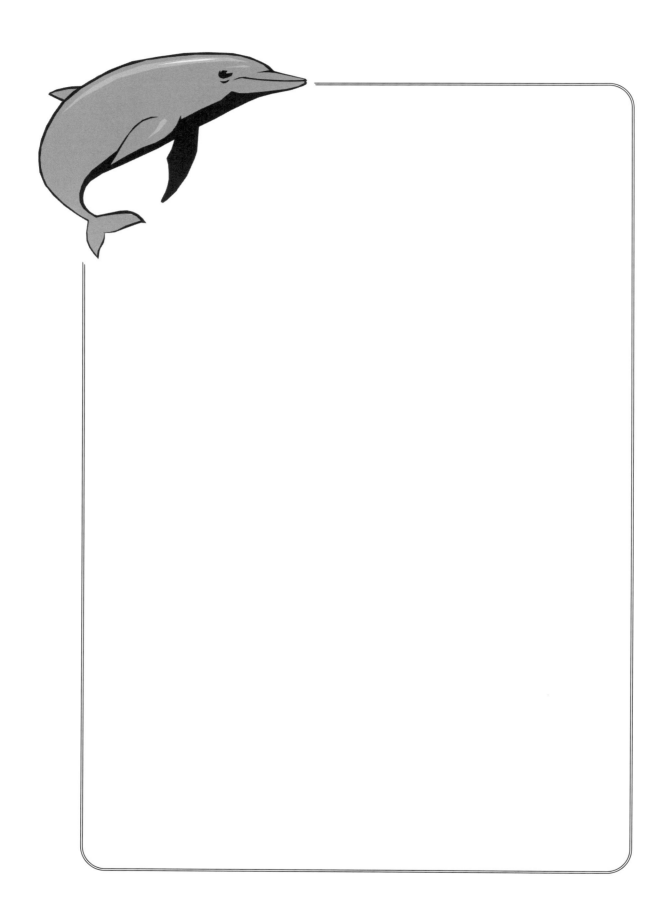

Trivia Questions

These questions have been included for fun and will reinforce some of the material that you might read during this study. Enjoy the search for answers, and then compare them with the answers that we found, located on the page 83.

1. What American statesman developed the first chart of the Gulf Stream, in response to delays in receiving mail from Britain?

2. Which ocean is the largest in the world?

3. What area of the globe is known as the Ring of Fire?

4. What does "scuba" stand for?

5. What kind of fish can be found living among sea anemones, without being harmed?

6. What type of sea animal is know for turning its stomach "inside-out" to eat?

7. What does "sonar" stand for?

8. What is the most plentiful form of plant life in the ocean?

9. Who is known as the "founder of oceanography" as well as the "Father of Annapolis"?

10. In the late 1800's, a new kind of communication innovation brought about a new era of learning about the ocean. What innovation and event brought this about?

Trivia Answers

1. What American statesman developed the first chart of the Gulf Stream, in response to delays in receiving mail from Britain?

   Benjamin Franklin

2. Which ocean is the largest in the world?

   The Pacific Ocean

3. What area of the globe is known as the Ring of Fire?

   The Pacific rim, which holds many of the world's volcanoes

4. What does "scuba" stand for?

   Self-Contained Underwater Breathing Apparatus

5. What kind of fish can be found living among sea anemones, without being harmed?

   Clown Fish

6. What type of sea animal is know for turning its stomach "inside-out" to eat?

   Sea star or starfish

7. What does "sonar" stand for?

   Sound Navigation And Ranging

8. What is the most plentiful form of plant life in the ocean?

   Phytoplankton

9.   Who is known as the "founder of oceanography" as well as the "Father of Annapolis"?

Matthew Maury

10.   In the late 1800's, a new kind of communication innovation brought about a new era of learning about the ocean. What innovation and event brought this about?

The laying of the Transatlantic Telegraph Cable

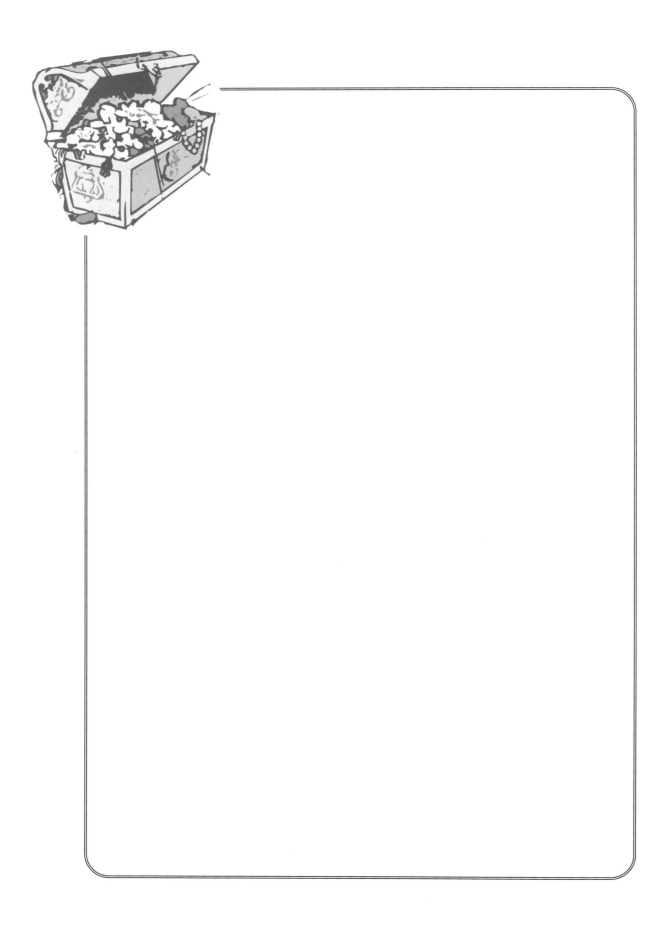

Reference Resources
The Oceans

Here is a list of books on the oceans to consider using as references for this unit study. Please understand that much of their content is useful, but that some contain evolutionary material. Look over the material and choose the portions that you feel meet with your requirements.

Our Amazing Ocean, by David Adler. (Question & Answer Books). Grades 3–6. Published by Troll Associates, 100 Corporate Dr., Mahwah, NJ 07430. (800) 929-8765.

Oceans & Seas, by Chris Arvetis and Carole Palmer. (Where Are We? Series). Grades 2–5. Published by Random House Books for Young Readers, 400 Hahn Rd., Westminster, MD 21157. (800) 733-3000.

Exploring the Ocean World: A History of Oceanography, by Clarence P. Idyll. An older book (© 1969), it provides a balanced and overall perspective on oceanography's history and scientists/explorers. Published by Thomas Y. Crowell Company, 900 S. Kansas, Topeka, KS 66612. (800) 899-9553 / (913) 232-9553.

Oceans, by Katharine Carter. (New True Books). Grades K–4. Published by Children's Press, P.O. Box 1331, Danbury, CT 06813. (800) 621-1115.

The Magic Schoolbus on the Ocean Floor, by Joanna Cole. Grades K–6. Published by Scholastic, Inc., P.O. Box 7502, Jefferson City, MO 65102. (800) 325-6149.

Ocean, by Miranda MacQuitty. (Eyewitness Books Series). Grades 4–12. Published by Alfred A. Knopf, subsidiary of Random House, 400 Hahn Rd., Westminster, MD 21157. (800) 733-3000. Available from Great Christian Books, 229 S. Bridge St., P.O. Box 8000, Elkton, MD 21922-8000. (800) 775-5422.

Ocean Facts, by B. Gibbs. (Usborne Facts & Lists Series). Grades 3–7. Published by EDC Publishing, 10302 E. 55th Place, Tulsa, OK 74146. (800) 475-4522.

Look Inside the Ocean, by Patrizia Malfatti. (Poke & Look Learning Series). Grades PreK–3. Published by Putnam Berkeley Group, 200 Madison Avenue, New York, NY 10016. (800) 631-8571.

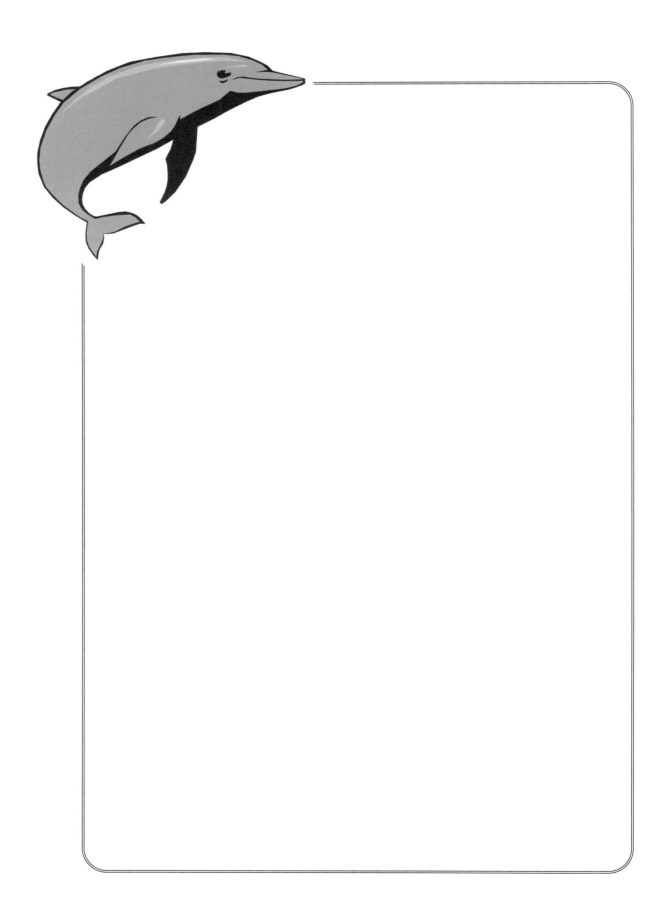

Seas & Oceans, by David Lambert. Grades 5–8. Published by Silver Burdett Press, Simon & Schuster, Inc., 200 Old Tappan Rd., Old Tappan, NJ 07675. (800) 223-2348.

The Sea, by Nina Morgan. (First Facts Series). Grades K–4. Published by Larousse, Kingfisher, Chambers, Inc., 95 Madison Avenue, New York, NY 10016. (800) 497-1657.

Oceans, by Callie Oldershaw. (Our Planet Series). Grades 4–6. Published by Troll Associates, 100 Corporate Dr., Mahwah, NJ 07430. (800) 526-5289.

Oceans, by Joy Palmer. (First Starts Series). Grades PreK–3. Published by Raintree Steck-Vaughn, P.O. Box 26015, Austin, TX 78755. (800) 531-5015.

Under the Sea, by Kate Petty. (Around & About Series). Grades 2–4. Published by Barron's Educational Series, Inc., 250 Wireless Boulevard, Hauppauge, NY 11788. (800) 645-3476.

Under The Sea, by Brenda Thompson and Cynthia Overbeck. (First Fact Books). Grades K–3. Published by Lerner Publications, 241 First Ave. N., Minneapolis, MN 55401. (800) 328-4929.

Ocean World, by Tony Rice. (Young Readers' Nature Library). Grades 4–6. Published by The Millbrook Press, 2 Old New Milford Rd., Brookfield, CT 06804-0335. (800) 462-4703.

Oceans, by Francene Sabin. Grades 3–6. Published by Troll Associates, 100 Corporate Dr., Mahwah, NJ 07430. (800) 929-8765.

The Oceans: A Book of Questions and Answers, by Don Groves. Grades 7–12. Published by John Wiley & Sons, Inc., 1 Wiley Drive, Somerset, NJ 08875 (800) 225-5945, ext. 2497.

Ocean: The Living World, by Barbara Taylor. Grades 1–7. Published by Dorling Kindersley. Distributed by Houghton Mifflin, Industrial/Trade Division, 181 Ballardvale Rd., Wilmington, MA 01887 (800) 225-3362.

World Around Us—The Sea, by Brian Williams. (World Around Us Series). Grades 3–8. Published by Larousse, Kingfisher, Chambers, Inc., 95 Madison Avenue, New York, NY 10016. (800) 497-1657.

Nature Hide & Seek: Oceans, by John Wood. Grades 1–4. Published by Knopf (subsidiary of Random House), 400 Hahn Rd., Westminster, MD 21157. (800) 733-3000.

Rivers & Oceans: Geography Facts & Experiments, by Barbara Taylor. (Young Discoverers Series). Grades 1–4. Published by Larousse, Kingfisher, Chambers, Inc., 95 Madison Avenue, New York, NY 10016. (800) 497-1657.

I Wonder Why the Sea is Salty, by Anita Ganeri. (Wonder Why Series). Grades 2–5. Published by Larousse, Kingfisher, Chambers, Inc., 95 Madison Avenue, New York, NY 10016. (800) 497-1657.

Powerful Waves, by D. M. Sousa. Grades 1–4. Published by Lerner Publications, 241 First Ave. N., Minneapolis, MN 55401. (800) 328-4929.

Tidal Waves & Other Ocean Wonders, by Q. L. Peace. (Amazing Science Series). Grades 4–6. Published by Silver Burdett, Simon & Schuster, Inc., 200 Old Tappan Rd., Old Tappan, NJ 07675. (800) 257-5755.

Reference Resources
History and Exploration

Here is a list of books on the history and exploration of the oceans to consider using as references for this unit study. Please understand that much of their content is useful, but that some contain evolutionary material. There are also too many biographies available concerning many of the famous explorers to list here, so use the list of famous explorers in the Subject Word List to locate more in your local library.

Unlocking the Mysteries of Creation, by Dennis R. Petermen. All ages. Published by Creation Resource Foundation, P.O. Box 570, El Dorado, CA 95623. (916) 626-4447.

The Amazing Story of Creation: From Science and the Bible, by Duane Gish. Grades 7 and up. Published by Master Books, P.O. Box 727, Green Forest, AR 72638. (800) 999-3777.

Noah's Ark and Lost World, by Dr. John Morris. Grades 2 and up. Published by Master Books, P.O. Box 727, Green Forest, AR 72638. (800) 999-3777. Available from Farm Country General Store, Rt. 1, Box 63, Metamora, IL 61548. (800) 551-FARM.

Explorer, by Rupert Matthews. (Eyewitness Books Series). Grades 4–12. Published by Alfred A. Knopf (subsidiary of Random House), 400 Hahn Rd., Westminster, MD 21157. (800) 733-3000.

Explorers, by F. Everett. (Usborne Famous Lives Series). Grades 4 and up. Published by EDC Publishing, 10302 E. 55th Place, Tulsa, OK 74146. (800) 475-4522.

Explorers, by Dennis Fradin. (New True Books Series). Grades K–4. Published by Children's Press, P.O. Box 1331, Danbury, CT 06813. (800) 621-1115.

Age of Exploration of the Milliken Transparency Reproducible Books Series. Grades 7–12. Published by the Milliken Publishing Company, P.O. Box 21579, St. Louis, MO 63132. (800) 325-4136.

Explorers and Mapmakers, by Peter Ryan. (Time Detective Series). Grades 4–7. Published by Dutton Children's Books, Division of Penguin USA, 375 Hudson St., New York, NY 10014. (212) 366-2000.

Explorers of the Undersea World, by Richard Gaines. (World Explorer Series). Grades 6–12. Published by Chelsea House, 1974 Sproul Rd., Suite 400, P.O. Box 914, Broomall, PA 19008. (800) 848-2665.

Viking from the Eyewitness Series. Grades 4–12. Published by Alfred A. Knopf (subsidiary of Random House), 400 Hahn Rd., Westminster, MD 21157. (800) 733-3000.

The Viking World from the Usborne Illustrated World History Series. Grades 5–10. Published by EDC Publishing, 10302 E. 55th Place, Tulsa, OK 74146. (800) 475-4522.

Christopher Columbus, by Bennie Rhodes. (The Sower Series). Grades 5 and up. Published by Mott Media, 1000 E. Huron, Milford, MI 48381. (800) 421-6645

Log of Christopher Columbus' First Voyage to America, by William Roy. Published by William Roy, P.O. Box 961, Taylors, SC 29687. Available from Great Christian Books, 229 S. Bridge St., P.O. Box 8000, Elkton, MD 21922-8000. (800) 775-5422.

Meet Christopher Columbus, by James DeKay. (Step-Up Biographies Series). Grades 2–4. Published by Random House Books for Young Readers, 400 Hahn Rd., Westminster, MD 21157. (800) 733-3000.

Where Do You Think You're Going, Christopher Columbus?, by Jean Fritz. Grades 3–7. Published by Putnam Berkely Group, 200 Madison Avenue, New York, NY 10016. (800) 631-8571.

The Last Crusader: The Untold Story of Christopher Columbus, by George Grant. Grades 8–12. Published by Crossway Books, 1300 Crescent Street, Wheaton, IL 60187. (708) 682-4300.

Explorations of Captain James Cook in the Pacific: As Told by Selections of His Own Journals, 1768-1779, by James Cook. Grades 9–12. Published by Dover Publications, 31 East 2nd Street, Mineola, NY 11501.

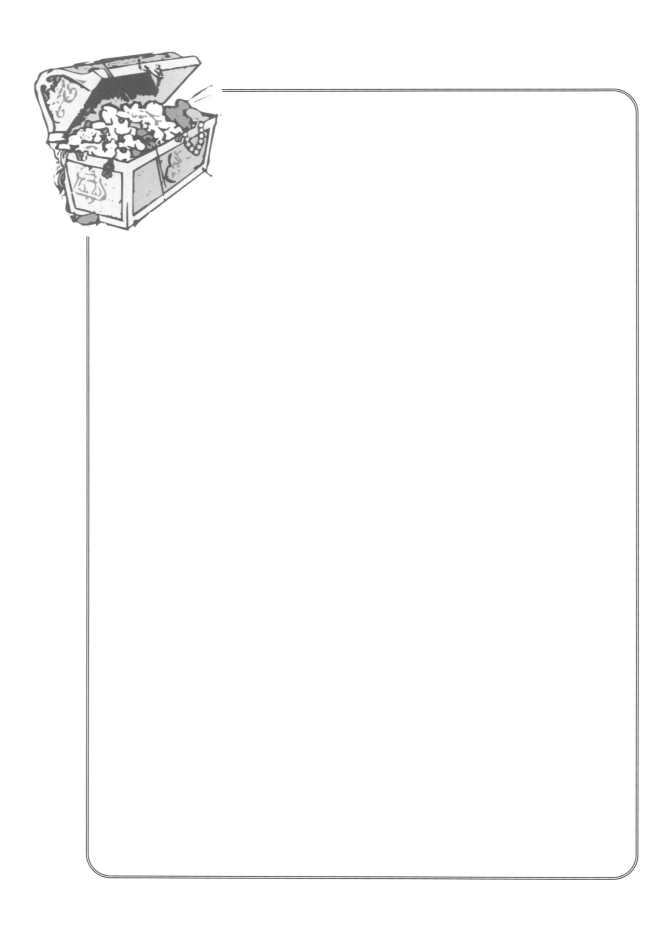

Carry On, Mr. Bowditch, by Jean L. Latham. Grades 5 and up. Published by Houghton Mifflin, Ind/Trade Division, 181 Ballardvale Rd., Wilmington, MA 01887 (800) 225-3362.

Robert Fulton: Boy Craftsman, by Henry. (Childhood of Famous Americans Series). Grades 2–6. Published by Aladdin Paperbacks, Simon & Schuster Children's Publishing Division, 200 Old Tappan Rd., Old Tappan, NJ 07675. (800) 257-5755.

Robert Fulton: Steamboat Builder, by Joann Landers-Henry. (Discovery Biography Series). Grades 2–6. Published by Chelsea House, 1974 Sproul Rd., Suite 400, P.O. Box 914, Broomall, PA 19008. (800) 848-2665.

Explorers of the Deep: Pioneers of Oceanography, by Donald W. Cox. Grades 6–12. An older book (© 1968), provides a broad view of oceanographic explorers, from Benjamin Franklin to modern explorers. Published by Hammond, Inc., Maplewood, NJ.

Deep Sea Explorer: The Story of Robert Ballard, Discoverer of the Titanic, by Rick Archbold. Grades 3–7. Published by Scholastic, Inc., P.O. Box 7502, Jefferson City, MO 65102. (800) 325-6149.

The Children's Atlas of Exploration by Antony Mason. Published by The Millbrook Press, 2 Old New Milford Rd., Brookfield, CT 06804-0335. (800) 462-4703.

Pirate, by Richard Platt. (Eyewitness Books Series). Grades 4–12. Published by Alfred A. Knopf (subsidiary of Random House), 400 Hahn Rd., Westminster, MD 21157 (800) 733-3000. Available from Great Christian Books, 229 S. Bridge St., P.O. Box 8000, Elkton, MD 21922-8000. (800) 775-5422.

Pirates and Patriots of the Revolution, by Keith Wilbur. Grades 4–8. Published by Globe Pequot Press, Box Q, Chester, CT 06412.

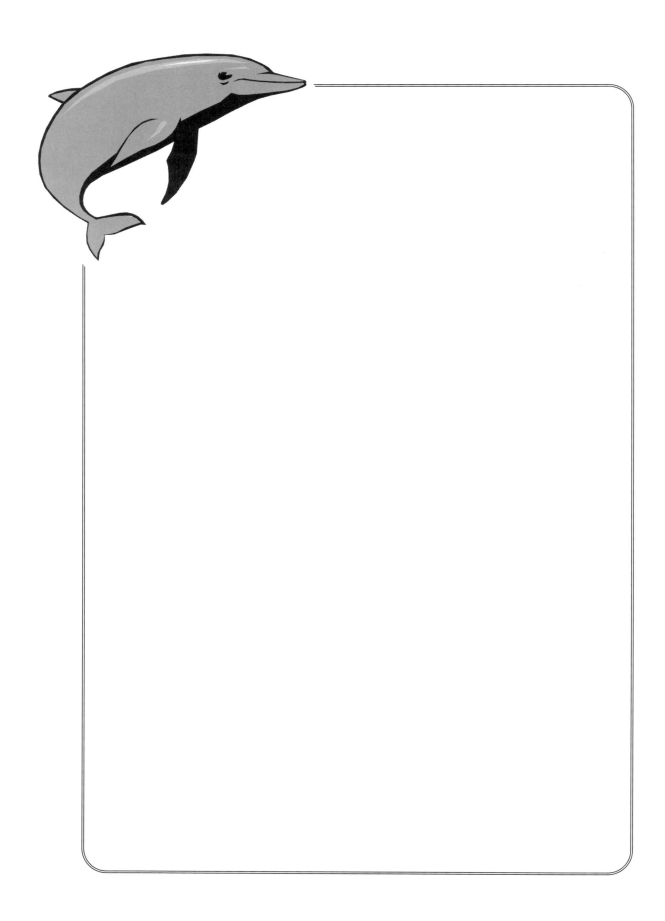

Reference Resources
Sea Life

Here is a list of books on the sea life in the oceans to consider using as references for this unit study. Please understand that much of their content is useful, but that some contain evolutionary material. Look over the material and choose the portions that you feel meet with your requirements.

Undersea, by C. Pick. (Young Scientist Series). Grades 1–5. Published by EDC Publishing, 10302 E. 55th Place, Tulsa, OK 74146. (800) 475-4522.

Sea Creatures. (What's Inside? Series). Grades K–3. Published by Dorling Kindersley. Distributed by Houghton Mifflin, Ind/Trade Division, 181 Ballardvale Rd., Wilmington, MA 01887 (800) 225-3362.

Fish, Shark and ***Whale***, all from the Eyewitness Books Series. Grades 4–12. Published by Alfred A. Knopf (subsidiary of Random House), 400 Hahn Rd., Westminster, MD 21157. (800) 733-3000.

How to Hide an Octopus: And Other Sea Creatures, by Ruth Heller. (All Aboard Books). Grades PreK–3. Published by Putnam Berkeley Group, 200 Madison Avenue, New York, NY 10016. (800) 631-8571.

Questions and Answers About: Sharks, by Ann McGovern. Grades 2–8. Published by Scholastic, Inc., P.O. Box 7502, Jefferson City, MO 65102. (800) 325-6149.

Underwater Life: The Oceans, by Dean Morris. Grades K–3. An older book (© 1977) that provides a good basic introduction to sea life for younger students. Published by Raintree Children's Books, Raintree Steck-Vaughn, PO Box 26015, Austin, TX 78755. (800) 531-5015.

What's In The Deep Blue Sea?, by Peter Seymour. (What's In The ... Series). Grades PreK–2. Published by Henry Holt & Co., 4419 West 1980 South, Salt Lake City, UT 84104. (800) 488-5233.

Wonders of the Sea, by Louis Sabin. Grades 2–4. Published by Troll Associates, 100 Corporate Dr., Mahwah, NJ 07430. (800) 929-8765.

Secrets of the Deep, by Ingrid Selberg. Grades 1–5. Published by Dial Books for Young Readers, Penguin USA, P.O. Box 120, Bergenfield, NJ 07621. (800) 253-6476 Consumer Sales.

What Lives in the Sea?, by Peter Seymour. Grades 2–5. Published by Simon & Schuster Children's, 200 Old Tappan Rd., Old Tappan, NJ 07675. (800) 257-5755.

My Visit to the Aquarium, by Aliki. Grades PreK–3. Published by HarperCollins Children's Books, 1000 Keystone Industrial Park, Scranton, PA 18512. (800) 242-7737.

Life in the Oceans, by Lucy Baker. Grades 4–7. Published by Scholastic, Inc., P.O. Box 7502, Jefferson City, MO 65102. (800) 325-6149.

A Picture Book of Underwater Life, by Theresa Grace. Grades 1–4. Published by Troll Associates, 100 Corporate Dr., Mahwah, NJ 07430. (800) 929-8765.

Let's Investigate Slippery, Splendid Sea Creatures, by Madelyn W. Carlisle. (Let's Investigate Series). Grades 3–7. Published by Barron's Educational Series, Inc., 250 Wireless Boulevard, Hauppauge, NY 11788. (800) 645-3476.

Corals: The Sea's Great Builders, by the Cousteau Society Staff. Grades 1–5. Published by Simon and Schuster Children's, 200 Old Tappan Rd., Old Tappan, NJ 07675. (800) 257-5755.

Hungry, Hungry Sharks and **Whales**, all titles from the Step Into Reading Series. Published by Random House Books for Young Readers, 400 Hahn Rd., Westminster, MD 21157. (800) 733-3000.

Dolphin, by Robert A. Morris. (Trophy I Can Read Book). Grades K–3. Published by HarperCollins Children's Books, 1000 Keystone Industrial Park, Scranton, PA 18512. (800) 242-7737.

Manatee: On Location, by Kathy Darling. Grades 4–7. Published by Lothrop, Lee and Shepard Books, William Morrow and Co., 39 Plymouth St., Fairfield, NJ 07004. (800) 237-0657.

Life In the Sea, by Eileen Curran. Grades K–2. Published by Troll Associates, 100 Corporate Dr., Mahwah, NJ 07430. (800) 929-8765.

Undersea, by Dan Mackie. (CHP Technology Series). Grades 4–8. Published by Durkin Hayes Publishing, 1 Colomba Drive, Niagara Falls, NY 14305. (800) 962-5200.

Under the Sea from A to Z, by Anne Doubilet. Grades K–6. Published by Crown Books for Young Readers, Division of Random House, 400 Hahn Rd., Westminster, MD 21157. (800) 726-0600.

Sharks and other Creatures of the Deep, by Philip Steele. (See & Explore Library). Grades 3 and up. Published by Dorling Kindersley. Distributed by Houghton Mifflin, Ind/Trade Division, 181 Ballardvale Rd., Wilmington, MA 01887 (800) 225-3362.

Don't Blink Now! Capturing the Hidden World of Sea Creatures, by Ann Downer. (New England Aquarium Books). Grades 5–8. Published by Franklin Watts, 5450 Cumberland Ave., Chicago, IL 60656. (800) 672-6672.

Tentacles: Octopus, Squid, and Their Relatives, by James Martin. Grades 2–6. Published by Crown Books for Young Readers, Division of Random House, 400 Hahn Rd., Westminster, MD 21157. (800) 726-0600.

Whales and Other Creatures of the Sea, by Joyce Milton. (Pictureback Series). Grades PreK–4. Published by Random Books for Young Readers, 400 Hahn Rd., Westminster, MD 21157. (800) 733-3000.

Whales, by Maura M. Gouck. (Nature Book Series). Grades 2–6. Published by Child's World, 505 Highway 169, Suite 295, Plymouth, MN 55441. (800) 599-7323.

Whales, by Judith Greenberg and Helen Carey. (Science Adventure Series). Grades 2–4. Published by Raintree Steck-Vaughn, P.O. Box 26015, Austin, TX 78755. (800) 531-5015.

Whale Watcher's Guide, by Robert Gardener. Grades 7–12. Published by Silver Burdett Press, Simon & Schuster, Inc., 200 Old Tappan Rd., Old Tappan, NJ 07675. (800) 257-5755.

Underwater Life, by Dean Morris. (Read About Series). Grades PreK–3. Published by Raintree Steck-Vaughn, P.O. Box 26015, Austin, TX 78755. (800) 531-5015.

Migration in the Sea, by Liz Oram and Robin Baker. (Migration Series). Grades 4–8. Published by Raintree Steck-Vaughn, P.O. Box 26015, Austin, TX 78755. (800) 531-5015.

Animals of the Sea and Shore, by Illa Podendorf. (New True Books). Grades K–4. Published by Children's Press, P.O. Box 1331, Danbury, CT 06813. (800) 621-1115.

Sponges are Skeletons, by Barbara Esbensen. (Let's-Read-&-Find-Out Series). Grades K–4. Published by HarperCollins Children's Books, 1000 Keystone Industrial Park, Scranton, PA 18512. (800) 242-7737.

Sea Jellies: Rainbows in the Sea, by Elizabeth Gowell. Grades 5–8. Published by Franklin Watts, 5450 Cumberland Ave., Chicago, IL 60656. (800) 672-6672.

Great Barrier Reef, by Martin Gutnik and Natalie Browne-Gutnik. Grades 4 and up. Published by Raintree Steck-Vaughn, P.O. Box 26015, Austin, TX 78755. (800) 531-5015.

Sam the Sea Cow, by Francine Jacobs. Grades 1–3. Published by Walker & Company, 435 Hudson St., New York, NY 10014. (800) 289-2553.

Manatees, by Emilie Lepthien. (New True Books). Grades K–4. Published by Children's Press, P.O. Box 1331, Danbury, CT 06813. (800) 621-1115.

How Did We Find Out About Life in the Deep Sea?, by Isaac Asimov. Grades 4–7. Published by Walker & Company, 435 Hudson St., New York, NY 10014. (800) 289-2553.

The Strange Eating Habits of Ocean Creatures, by Jean Sibbald. Grades 4–8. Published by Silver Burdett, Simon & Schuster, Inc., 200 Old Tappan Rd., Old Tappan, NJ 07675. (800) 223-2348.

The Marine Biology Coloring Book, by Thomas Nielsen. Grades 7–12. Published by HarperCollins, 1000 Keystone Industrial Park, Scranton, PA 18512. (800) 242-7737. Available from Farm Country General Store, Rt. 1, Box 63, Metamora, IL 61548. (800) 551-FARM.

The Seaside Naturalist: A Guide to Study at the Seashore, by Deborah A. Coulombe. Grades 4–12. Published by Prentice Hall Press, 200 Madison Ave., New York, NY 10016. (800) 631-8571.

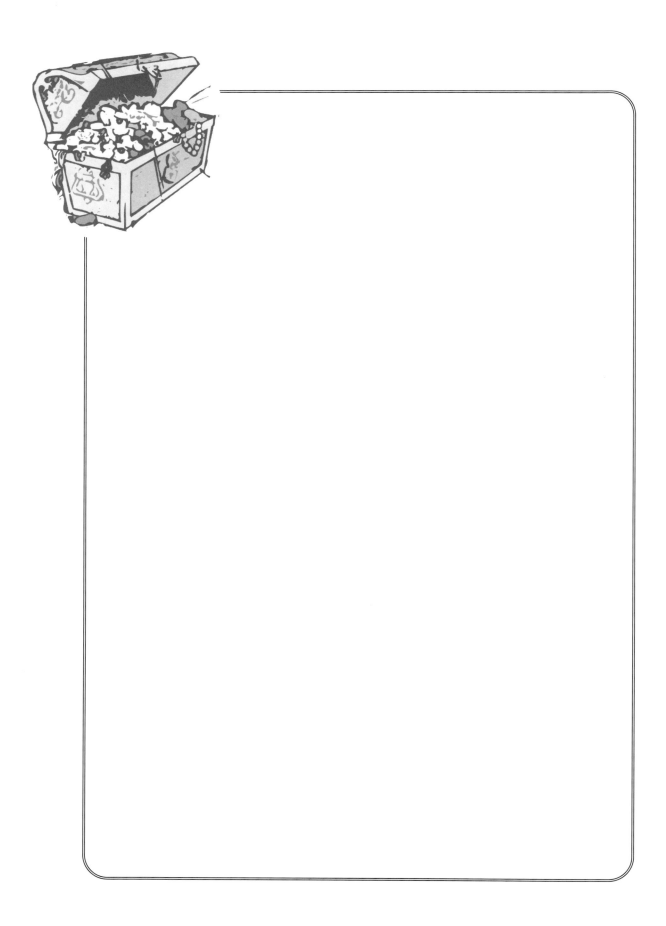

Way Down Deep, by Patricia B. Demuth. (All Aboard Reading Series). Grades 1–3. Published by Putnam Berkeley Group, 200 Madison Avenue, New York, NY 10016. (800) 631-8571.

Deep Sea Vents: Living Worlds Without Sun, by John F. Waters. Grades 5 and up. Published by Dutton Children's Books, Division of Penguin USA, 375 Hudson St., New York, NY 10014. (800) 253-6476.

Reference Resources
Geography

Here is a list of books on topics related to geography of the oceans, to consider using as references for this unit study. Please understand that much of their content is useful, but that some contain evolutionary material. Look over the material and choose the portions that you feel meet with your requirements.

Seas and Oceans, by Felicity Brooks. (Understanding Geography Series). Grades 5–9. Published by Usborne, EDC Publishing, 10302 E. 55th Place, Tulsa, OK 74146. (800) 475-4522.

Janice Van Cleave's Geography for Every Kid, by Janice Van Cleave. Grades 4 and up. Published by John Wiley & Sons, 1 Wiley Dr., Somerset, NJ 08875. (800) 225-5945, ext. 2497.

The Oceans Atlas, by Anita Ganeri. Grades 4–12. Published by Dorling Kindersley. Distributed by Houghton Mifflin, Ind/Trade Division, Ballardvale Rd., Wilmington, MA 01887 (800) 225-3362.

Maps: Getting From Here to There, by Harvey Weiss. Grades 3–8. Published by Houghton Mifflin Company, Ind/Trade Division, Ballardvale Rd., Wilmington, MA 01887 (800) 225-3362.

Arctic and Antarctic (Eyewitness Series). Grades 4–12. Published by Alfred A. Knopf (subsidiary of Random House), 400 Hahn Rd., Westminster, MD 21157. (800) 733-3000.

Oceans and Seas, by Chris Arvetis and Carole Palmer. (Where Are We? Series). Grades PreK–4. Published by Rand McNally, Attn: World Atlas, 8255 N. Central Park, Skokie, IL 60076. (800) 333-0136. Available from Great Christian Books, 229 S. Bridge St., P.O. Box 8000, Elkton, MD 21922-8000. (800) 775-5422.

Mapping The World By Heart, by David Smith. Grades 7–12. Published by Tom Snyder Productions, 80 Coolidge Hill Rd., Watertown, MA 02172. (800) 342-0236.

The Geography Coloring Book, by Wynn Kapit. Grades 7–12. Published by HarperCollins Publishers, 1000 Keystone Industrial Park, Scranton, PA 18512. (800) 328-3443.

Reference Resources
Seashore

Here is a list of books on topics related to the seashore to consider using as references for this unit study. Please understand that much of their content is useful, but that some contain evolutionary material. Look over the material and choose the portions that you feel meet with your requirements.

Exploring an Ocean Tide Pool, by Jeanne Bendick. Grades 2–4. Published by Henry Holt & Co., 4419 West, 1980 South, Salt Lake City, UT 84104. (800) 488-5233.

The Seaside Naturalist: A Guide to Study at the Seashore, by Deborah A. Coulombe. Grades 4–12. Published by Prentice Hall Press, Simon & Schuster, Attn: Order, 200 Old Tappan Rd., Old Tappan, NJ 07675. (800) 223-2348.

Coral Reef, Shoreline and *Tide Pool* from the Look Closer Series. Grades PreK–3. Published by Houghton Mifflin Co., Ind./Trade Division, 181 Ballardvale Rd., Wilmington, MA 01887 (800) 225-3362.

A Walk on the Great Barrier Reef, by Caroline Arnold. Grades 3–8. Published by Carolrhoda Books, Inc. (division of Lerner Publications), 241 First Ave., N., Minneapolis, MN 55401. (800) 328-4929.

Seashells, by Carol Benanti. (Collector's Kits Series). Grades 3 and up. Published by Random House Books for Young Readers, 400 Hahn Rd., Westminster, MD 21157. (800) 733-3000.

Seashore Surprises, by Rose Wyler. (Outdoor Science Series). Grades K–3. Published by Silver Burdett, Simon & Schuster, Inc., 200 Old Tappan Rd., Old Tappan, NJ 07675. (800) 223-2348.

Seashells of North America, by R. Tucker Abbott. (Golden Field Guide Series). Grades 9–12. Published by Western Publishing, 1220 Mound Ave., Racine, WI 53404. (888) 732-3263.

What Lives In a Shell?, by Kathleen W. Zoehfeld. (Let's Read and Find-Out Science Series). Grades PreK–2. Published by HarperTrophy, 1000 Keystone Industrial Park, Scranton, PA 18512. (800) 242-7737.

Shell and Seashore, both from the Eyewitness Books Series. Grades 4–12. Published by Alfred A. Knopf (subsidiary of Random House), 400 Hahn Rd., Westminster, MD 21157. (800) 733-3000.

Animals of the Sea and Shore, by Illa Podendorf. (New True Books). Grades K–4. Published by Children's Press, P.O. Box 1331, Danbury, CT 06813. (800) 621-1115.

The Seashore, by Jane Walker and David Marshall. (Fascinating Facts Series). Grades 3–4. Published by Millbrook Press, 2 Old New Milford Rd., Brookfield, CT 06804-0335. (800) 462-4703.

America's Lighthouses: An Illustrated History, by Francis R. Holland, Jr. Published by Dover Publications, 31 East 2nd Street, Mineola, NY 11501.

Reference Resources
Careers

A Day in the Life of a Marine Biologist, by William Jaspersohn. Grades 5–12. Published by Little, Brown & Co., 200 West St., Waltham, MA 02154. (800) 759-0190.

I Can Be an Oceanographer, by Paul Sipiera. (I Can Be Series). Grades K–3. Published by Children's Press, P.O. Box 1331, Danbury, CT 06813. (800) 621-1115.

A Day in the Life of a Marine Biologist, by David Paige. Grades 4–8. Published by Troll Associates, 100 Corporate Dr., Mahwah, NJ 07430. (800) 929-8765.

The Complete New Guide to Environmental Careers, by The Environmental Careers Organization. Grades 9–12. Published by Island Press, P.O. Box 7, Covelo, CA 95428. (800) 828-1302.

Opportunities in Marine and Maritime Careers, by William Ray Heitzmann. Grades 9–12. Published by VGM Career Horizons, a division of NTC Publishing Group, 4255 West Touhy Ave., Lincolnwood, IL 60646.

Jobs in Marine Science, by Frank Ross. (Exploring Careers Series). Grades 5 & up. Published by LOTHROP, Lee and Shephard Books, WIlliam Morrow and Co., P.O. Box 1219, Fairfield, NJ 07007. (800) 237-0657.

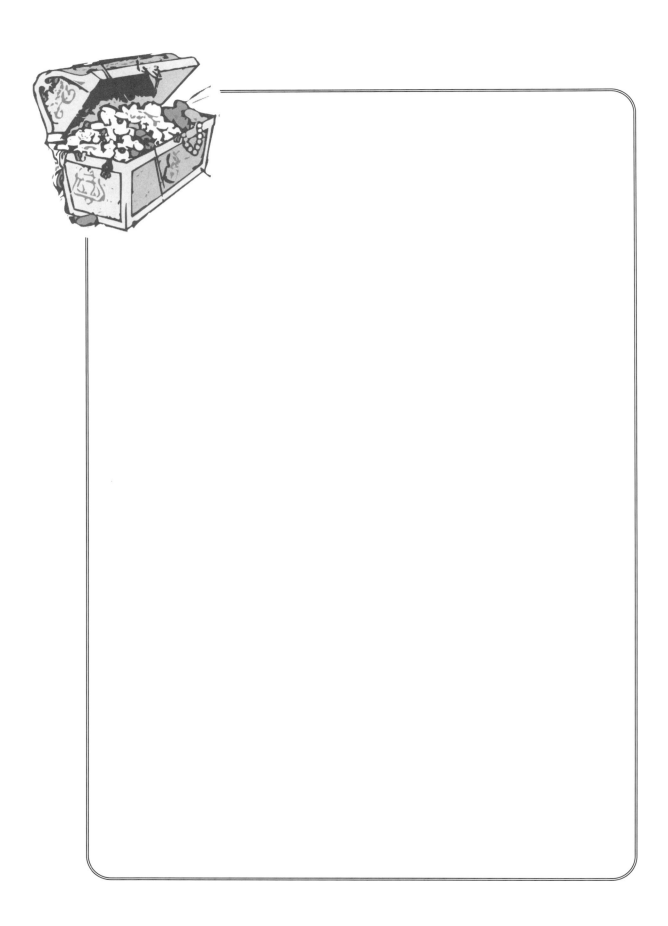

Reference Resources
Art

Draw Science: Whales, Sharks, and Other Sea Creatures, by Nina Kidd. Grades 5–12. Published by Lowell House Juvenile, 2029 Century Park East, Suite 3290, Los Angeles, CA 90067.

Draw 50 Boats, Ships, Trucks and Trains and **Draw 50 Sharks, Whales and Other Sea Creatures**, both by Lee J. Ames. Published by Doubleday & Co., Consumer Services, P.O. Box 507, Des Plaines, IL 60017. (800) 223-6834.

The Tall Ships of Today in Photographs, by Frank O. Braynard. Published by Dover Publications, 31 East 2nd Street, Mineola, NY 11501.

Life on a Fishing Boat: A Sketchbook, by Huck Scarry. Grades 3 and up. Published by Prentice-Hall, Inc., c/o Simon & Schuster, 200 Old Tappan Rd., Old Tappan, NJ 07675. (800) 223-2348.

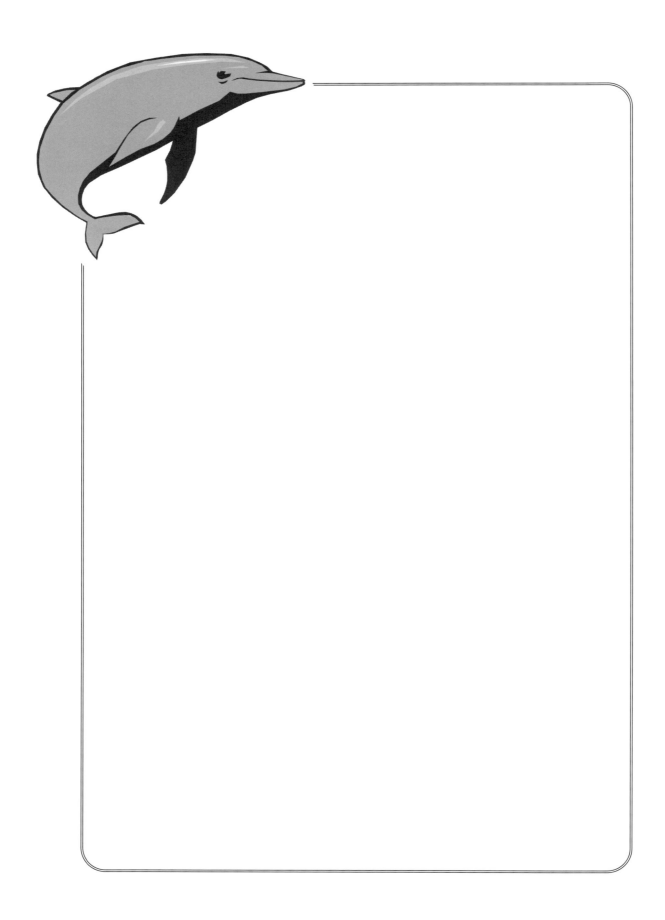

Reference Resources
Ships

The Titanic, (Step Into Reading Series). Grades 2–4. Published by Random House Books for Young Readers, 400 Hahn Rd., Westminster, MD 21157. (800) 733-3000.

Boat and ***Explorer***, both of the Eyewitness Books Series. Grades 4–12. Published by Alfred A. Knopf (subsidiary of Random House), 400 Hahn Rd., Westminster, MD 21157. (800) 733-3000.

Ships: Crossing the World's Oceans, by Sean M. Grady. (Encyclopedia of Discovery & Invention Series). Grades 5–8. Published by Lucent Books, Inc., P.O. Box 289011, San Diego, CA 92198-0011.

Amazing Boats, by Margaret Lincoln. (Eyewitness Junior Series). Grades 1–5. Published by Alfred A. Knopf (subsidiary of Random House), 400 Hahn Rd., Westminster, MD 21157. (800) 733-3000.

Boats, Ships, Submarines and Other Floating Machines, by Ian Graham. (How Things Work Series). Grades 3–7. Published by Larousse, Kingfisher, Chambers, Inc. Books, 95 Madison Ave., New York, NY 10016. (800) 497-1657.

Life on a Fishing Boat: A Sketchbook, by Huck Scarry. Grades 3 and up. Published by Prentice-Hall, Inc., c/o Simon & Schuster, 200 Old Tappan Rd., Old Tappan, NJ 07675. (800) 223-2348.

See Inside: A Submarine, by J. Rutland. (See Inside Series). Grades 2–8. Published by Warwick Press, 387 Park Avenue, New York, NY 10016.

Sunken Treasure, by Gail Gibbons. Grades 1–5. Published by Harper Trophy, Division of HarperCollins, 1000 Keystone Industrial Park, Scranton, PA 18512. (800) 242-7737.

Shipwrecks: A 3-Dimensional Exploration, by David Hawcock and Garry Walton. Grades 3 and up. Published by Harper Festival, a division of HarperCollins Publishers, 1000 Keystone Industrial Park, Scranton, PA 18512. (800) 242-7737.

Book of Old Ships: From Egyptian Galleys to Clipper Ships, by Henry B. Culver. Grades 8–12. Published by Dover Publications, 31 East 2nd Street, Mineola, NY 11501.

Visual Dictionary of Ships and Sailing (Eyewitness Books) Grades 4–12. Published by Dorling Kindersley, Houghton Mifflin, Ind./Trade Division, 181 Ballardvale Rd., Wilmington, MA 01887 (800) 225-3362. Available from Great Christian Books, 229 S. Bridge St., P.O. Box 8000, Elkton, MD 21922-8000. (800) 775-5422.

Ships: Sailors and the Sea, by Humble. Grades 5 and up. Published by Franklin Watts, 5450 Cumberland Ave., Chicago, IL 60656. (800) 672-6672.

American Sailing Ships: Their Plans and History, by Charles G. Davis. Grades 9–12. Published by Dover Publications, 31 East 2nd Street, Mineola, NY 11501.

Donald McKay and His Famous Sailing Ships, by Richard C. McKay. Grades 7–12. Published by Dover Publications, 31 East 2nd Street, Mineola, NY 11501.

Narratives of the Wreck of the Whale-Ship Essex, by Owen Chase. Eyewitness accounts of the incident which inspired Melville's ***Moby Dick***. Grades 9 and up. Published by Dover Publications, 31 East 2nd Street, Mineola, NY 11501.

Reference Resources
Miscellaneous

Record Breakers of the Sea, by Rupert Matthews. Grades 2–6. Published by Troll Associates, 100 Corporate Dr., Mahwah, NJ 07430. (800) 929-8765.

Diving Into Oceans, by the National Wildlife Federation Staff. Grades K–8. Published by National Wildlife Federation, Order Dept., P.O. Box 8925, Vienna, VA 22183. (800) 432-6564.

Life In The Great Ice Age, by Michael and Beverly Oard. Grades 4–12. Published by Master Books, P.O. Box 727, Green Forest, AR 72638. (800) 999-3777. Available from Farm Country General Store, Rt. 1, Box 63, Metamora, IL 61548. (800) 551-FARM.

Hurricanes and Typhoons, by Jacqueline Dineen. (Natural Disasters Series). Grades 3–8. Published by Shooting Star Press, 230 Fifth Avenue, New York, NY 10001.

Storm Warnings: Tornadoes and Hurricanes, by Jonathan D. Kahl. Grades 5–9. Published by Lerner Publications, 241 First Ave. N., Minneapolis, MN 55401. (800) 328-4929.

Magic Schoolbus in the Eye of the Hurricane, by Joanna Cole. Grades 1–4. Published by Scholastic, Inc., P.O. Box 7502, Jefferson City, MO 65102. (800) 325-6149.

Reading Resources

All of the following classics are available in various editions from several different publishers. The original versions are fascinating to read together as a family, while there are some abridged versions that might be more understandable to younger children.

Swiss Family Robinson, by Johann Wyss
20,000 Leagues Under the Sea, by Jules Verne
Robinson Crusoe, by Daniel Defoe
Treasure Island, by Robert Louis Stevenson
Kidnapped, by Robert Louis Stevenson
Moby Dick, by Herman Melville
The Old Man and the Sea, by Ernest Hemingway

Other books to consider for enjoyable reading during this unit study include:

Kon Tiki, by Thor Heyerdahl. Grades 8–12. Published by Simon & Schuster Trade, 200 Old Tappan Rd., Old Tappan, NJ 07675. (800) 223-2348.

Trapped at the Bottom of the Sea, by Frank Peretti. (Cooper Kids Adventures Series). Grades 4–7. Published by Crossway Books, Division of Good News Publications, 1300 Crescent St., Wheaton, IL 60187. (800) 635-7993.

Island of the Blue Dolphin, by Scott O'Dell. Grades 3–7. A Dell Yearling Book, published by Bantam, Doubleday, Dell, 2451 S. Wolf Rd., Des Plains, IL 60018. (800) 323-9872.

Pagoo, Paddle to the Sea and **Seabird**, all by Holling C. Holling. Grades 3–9. Published by Houghton Mifflin Co., Ind./Trade Division, 181 Ballardvale Rd., Wilmington, MA 01887 (800) 225-3362.

Noah's Ark, by Linda Hayward. (Step Into Reading Series). Grades PreK–1. Published by Random House Books for Young Readers, 400 Hahn Rd., Westminster, MD 21157. (800) 733-3000.

Mr. Popper's Penguins, by Richard Atwater. Grades 3–7. Published by Little Brown and Company, 200 West St., Waltham, MA 02154. (800) 759-0190.

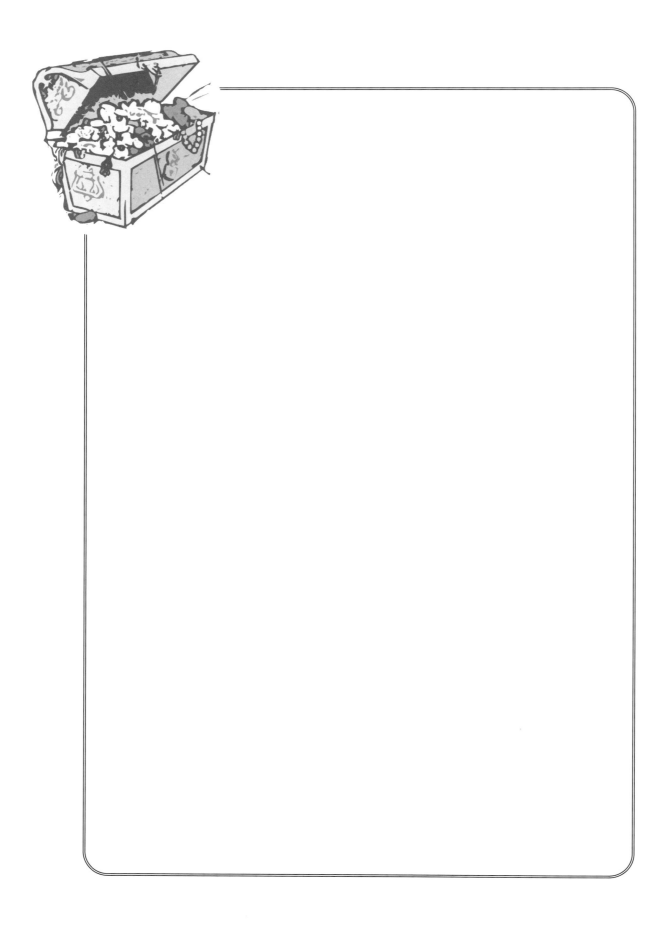

Keep the Lights Burning Abbie, by Peter and Connie Roop. Grades K–3. Published by Lerner Publications, 241 First Ave. N., Minneapolis, MN 55401. (800) 328-4929. This book, along with the Progeny Press Bible-based Literature Guide, are available from The Elijah Company, Route 2 Box 100-B, Crossville, TN 38555. (615) 456-6284.

Sea Squares, by Joy N. Hulme. Grades PreK–3. Published by Hyperion Books for Children, 114 Fifth Avenue, New York, NY 10011.

Seashore Book, by Charlotte Zolotow. Grades PreK–3. Published by HarperCollins Children's Books, 1000 Keystone Industrial Park, Scranton, PA 18512. (800) 242-7737.

Sailing to the Sea, by Mary Claire Helldorfer. Grades PreK–3. Published by Puffin Books, Penguin USA, P.O. Box 120, Bergenfield, NJ 07621. (800) 253-6467.

Lighthouse Mystery, by Gertrude C. Warner. (Boxcar Children Mysteries Series). Grades 2–7. Published by Albert Whitman and Co., 6340 Oakton St., Morton Grove, IL 60053. (800) 255-7675.

Whalesong, by Robert Siegel. Grades 8–12. Published by Crossway Books (division of Good News Publishers), 1300 Crescent Street, Wheaton, IL 60187. (800) 635-7993.

Working Outline

I. **Introduction**

    A.   Definition and origin of the words "ocean" and "sea"

    B.   Importance of studying the ocean

        1.   Vast resources for the future

        2.   Keys to our future as well as the past

        3.   Understanding the world and the delicate balance that God created

II. **History and Exploration**

    A.   Creation

        1.   "In the beginning God created the heaven and the earth." Genesis 1:1 (KJV)

        2.   "And God called the dry land Earth; and the gathering together of the waters called he Seas: and God saw that it was good." Genesis 1:10 (KJV)

3. "And God said, Let the waters bring forth abundantly the moving creature that hath life..." Genesis 1:20 (KJV)

4. "And God created great whales, and every living creature that moveth, which the waters brought forth abundantly, after their kind, and every winged fowl after his kind: and God saw that it was good." Genesis 1:21 (KJV)

5. "For in six days the Lord made heaven and earth, the sea, and all that in them is..." Exodus 20:11 (KJV)

B. Exploration

1. Early travels

   a. Primitive boats

   b. Early routes of water travel

2. Early explorers

   a. Phoenicians

   b. Greeks

c. Romans

d. Norsemen (Vikings)

3. Explorers in the Age of Discovery

   a. Henry the Navigator (Prince Henry) of Portugal

   b. Christopher Columbus

   c. Vasco da Gama

   d. Bartholomew Diaz

   e. John Cabot

   f. Vasco Nuñez de Balboa

   g. Juan Ponce de León

h. Hernando Cortez

i. Ferdinand Magellan

j. Francisco Vasquez de Coronado

k. Hernando de Soto

l. Francis Drake

m. Walter Raleigh

4. Later ocean explorers

a. Abel Tasman

b. James Cook

c. Jacques Piccard

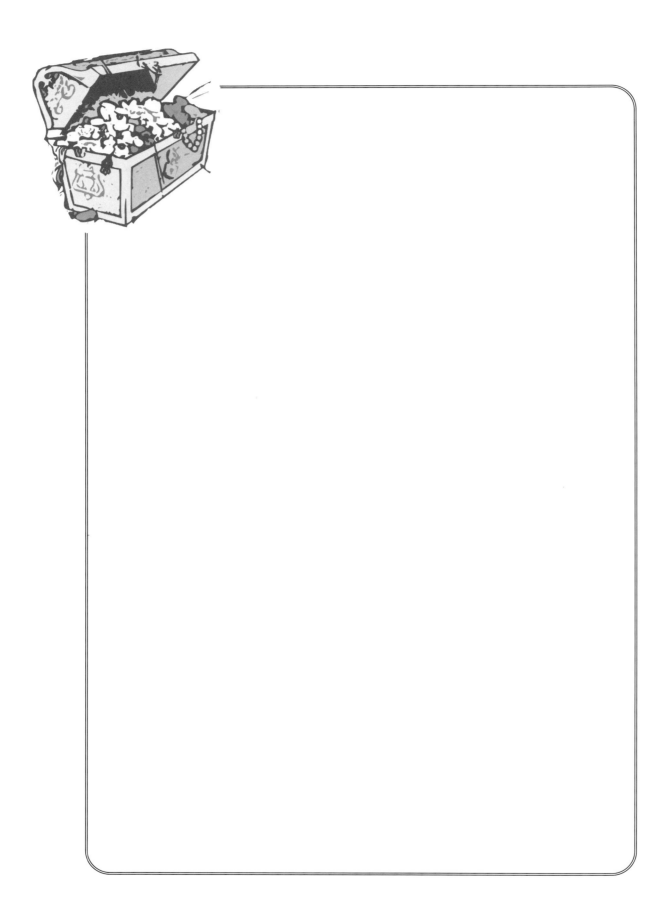

d.   Donald Walsh

e.   Jacques-Yves Cousteau

III. Geography

A.   Earth

1.   Dimensions

a.   Circumference

b.   Diameter

c.   Distance from the sun

d.   Volume

e.   Mass

2.   More than 70% of the Earth's surface is covered by water

3. Magnetic poles

B. Oceans

1. Major oceans

a. Atlantic Ocean *37,828,000 miles depth 12,1000 deepest point 29,010 where Puerto Rico Trench, location - Between east coasts of north & S. America & W. coasts, of Europe & africa*

b. Pacific Ocean *located - Betw. east coasts of asia & australia & W. coasts of N & S America depth - 13,355 - The deepest challenge deep 36,200*

c. Indian Ocean *- located - S. of asia betw. the East coast of africa, the W. coast of australia avg. depth - 10,037 - the deepest point Java trench - 25,344*

d. Arctic Ocean *- Surrounds the n. pole has 3,952 sq. ft. the deepest is in Eurasia Basin - 17,880*

e. Antarctic Ocean *- Surrounds the South pole or antarctic continent avg. 12,240 deepest 21,038*

2. Continental shelf

a. Components

b. Geographical features

3.   Geographical features of the ocean floor

     a.   Plates

     b.   Mountain ranges

     c.   Abyssal basins

     d.   Abyssal plains

     e.   Seamounts

     f.   Trenches

     g.   Fracture zones

4.   Islands

     a.   Atolls

     b.   Volcanic activity

      c.   Major island groups

   5.   Other geographical features

      a.   Coral reefs

      b.   Deep sea vents

C.   Continents

   1.   North America

   2.   South America

   3.   Europe

   4.   Asia

   5.   Australia

   6.   Antarctica

7. Africa

D. Maps

    1. Uses

        a. Direction (North, South, East, West)

        b. Location

            (1) Lines of latitude

                (a) Horizontal

                (b) Includes the equator

            (2) Lines of longitude

                (a) Vertical - from North to South Poles

                (b) Includes the Greenwich Meridian

        c. Surface characteristics

            (1) Topography

            (2) Geographical characteristics

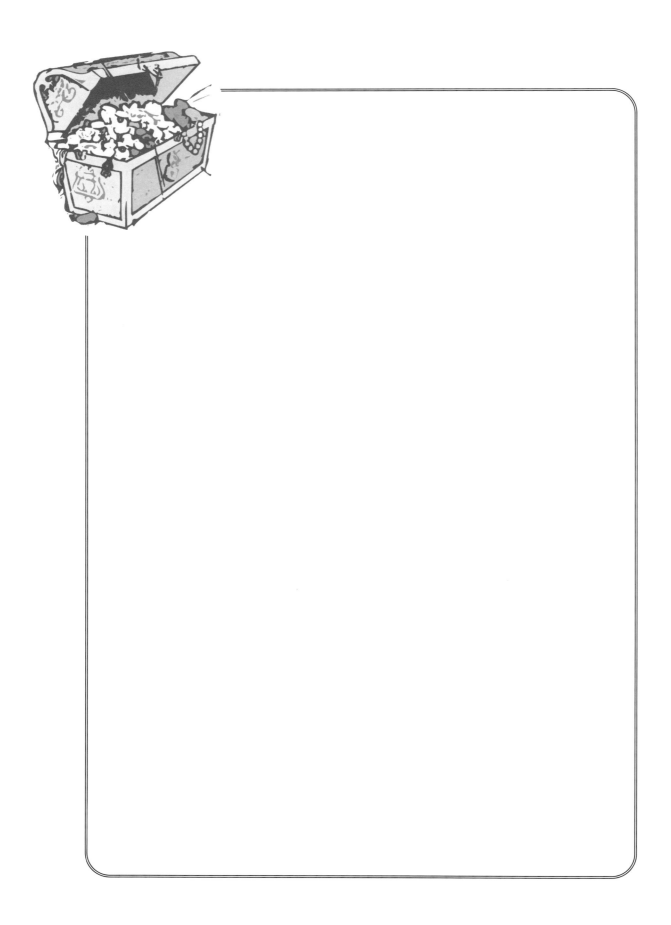

(3) Other features of an area (for example: population, rainfall amounts, ocean currents, etc.)

2. Features of a map

    a. Scale

    b. Legend

    c. Indicator of the direction of North

IV. Oceanography

A. Movement of the water

1. Currents

    a. Types of currents

        (1) Surface currents

        (2) Subsurface currents

    b. Causes of currents

        (1) Coriolis force

        (2) Wind on the surface

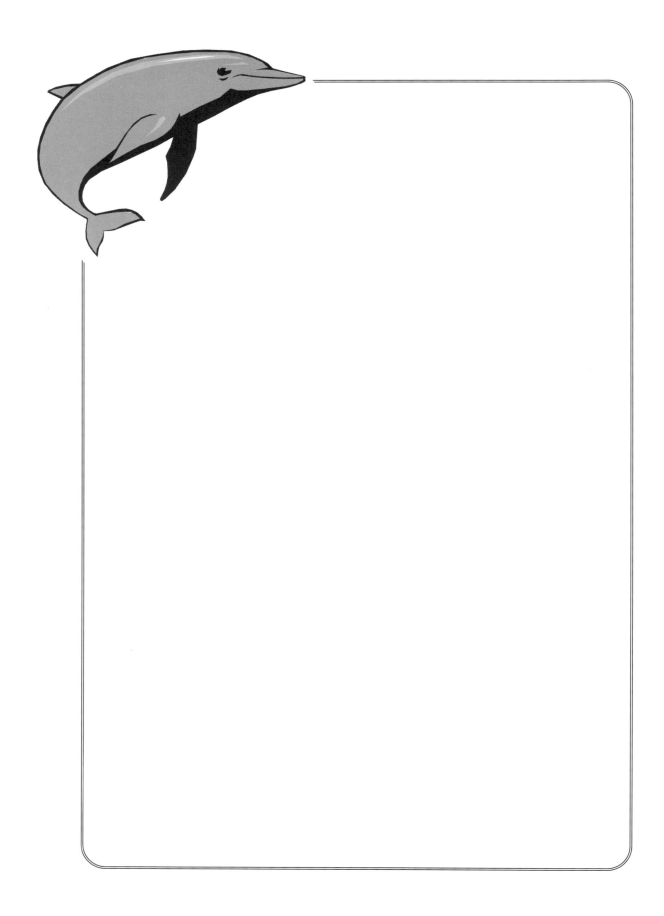

(3)　　Water density differences

2.　　Tides

    a.　　Tidal cycle

        (1)　　High tide

        (2)　　Low tide

    b.　　Astronomical tides

        (1)　　Neap tide

            (a)　　Lower-than-normal high tide

            (b)　　Occur when sun and moon are at right angles

        (2)　　Spring tide

            (a)　　Higher-than-normal high tide

            (b)　　Occur when sun and moon are lined up with the Earth in a straight line

3.　　Waves

    a.　　Causes of waves

(1)   Wind acting on the surface of the ocean

(2)   Sudden geologic changes under the surface of the ocean—volcanic eruption, earthquake, etc.

b.   Wave description terms

(1)   Crest (top)

(2)   Trough (bottom)

(3)   Height (distance from crest to trough)

(4)   Wavelength (distance between waves)

B.   Chemistry of the ocean water

1.   Chemical components

a.   Water (hydrogen and oxygen)

b.   Sodium

c.   Magnesium

d.   Calcium

e.   Potassium

        f.   Sulfates

        g.   Chlorides

    2.   Salinity of the ocean water

    C.   Products from the ocean

        1.   Food

            a.   Fish

            b.   Plants (seaweed, etc)

        2.   Petroleum products

        3.   Minerals (manganese nodules, for example)

V.   Ships

    A.   History

        1.   Reasons for building ships

            a.   To explore

b. To develop trade

c. To find other people

2. Shipbuilders in history

a. Phoenicians

b. Vikings

c. Greeks

d. Romans

e. Chinese

B. Types

1. Raft

2. Sail-propelled

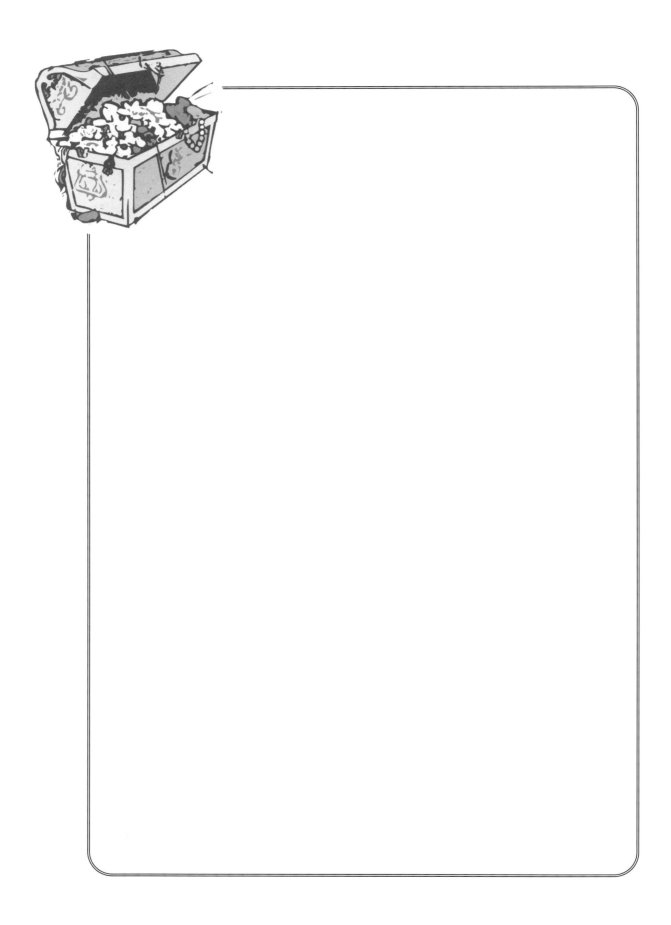

3. Human-propelled

4. Engine-propelled

5. Remote-controlled submersible

C. Study of maritime history

　　1. Meaning of maritime

　　2. Impact of ships on history

D. Famous shipwrecks

　　1. *Nuestra Señora de Atocha*

　　2. *Mary Rose*

　　3. *Titanic*

　　4. *Lusitania*

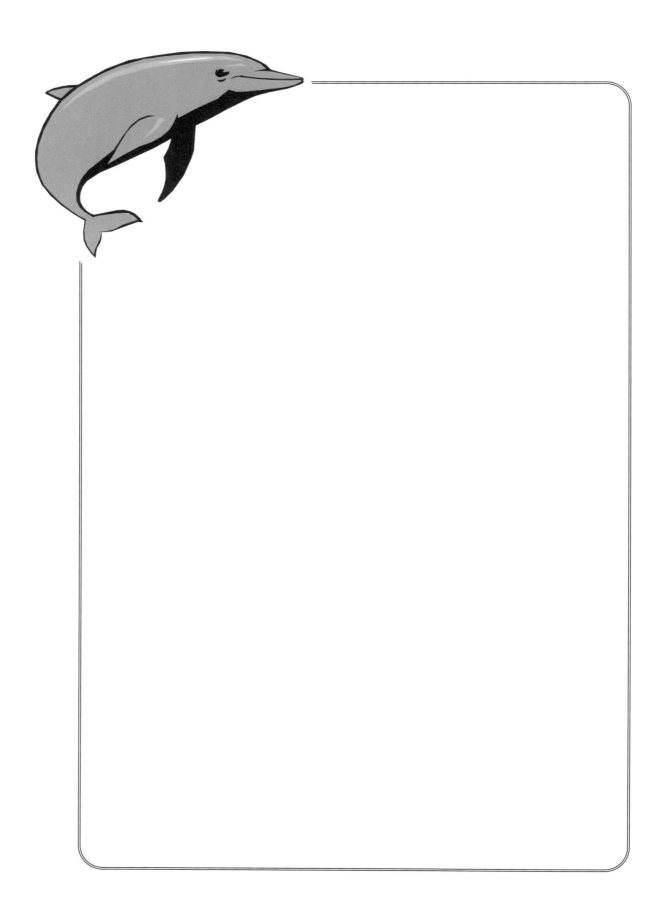

5. *Bismark*

6. *S. S. Central America*

VI. Navigation

A. Latin origin of the word "navigate"

B. Four basic methods of marine navigation

    1. Celestial navigation—using the sun, moon and stars

    2. Piloting navigation—using visual landmarks and depth soundings

    3. Dead reckoning navigation—using a carefully-maintained record (or "reckoning") of the ship's motion (time, direction and distance)

    4. Electronic navigation—using radar, radio direction finder, etc.

C. Tools of navigation

    1. Early navigational tools

         a.   Astrolabe

         b.   Sextant

    2.   Compass

    3.   Nautical charts

    4.   Dividers

    5.   Navigational lights, markers and buoys

VII. Marine biology (sea life)

  A.   Three "zones" of life in the ocean

    1.   Sunlit zone

    2.   Twilight zone

    3.   Bathypelagic zone

B.   Plant life examples

    1.   Phytoplankton

    2.   Seaweed

C.   Animal life examples [due tues].

    (1.)   Whale - Biggest whale - Blue whale its 100 ft long
            lives in the cold waters of north & South america
            weigh - 200 tons.

    (2.)   Dolphin - Dolphins are not porpoise!
            lives up to 20 yrs. - 12 ft long weigh 400 lbs.
            frequent warm coastal waters

    3.   Sea walrus

    (4.)   Seal Seals are graceful swimmers but are clumsy walkers

    5.   Tuna

    6.   Marlin

    7.   Swordfish

? (8.) Snapper

9. Bonito

(10.) Shark - whale Shark - 45ft long. - 65-70ft long.
found in tropical - sub tropical seas
weighs 25 tons

11. Sponge

12. Coral

13. Sea turtle

(14.) Jellyfish - is a name used to describe 200 different
kinds of ocean animals w/ delicate jellylike bodies.
they live throughout the ocean world. from icy
waters to the more warm waters. they have umbrella
shapes, they measure betw. 1 to 16 inches across
some are poisness.

(15.) Eel
there are more than 500 species of eels
66ft weigh about 110 lbs
live along the Atlantic coasts of both Europe, N. America.

(16.) Starfish

(17.) Sea urchin

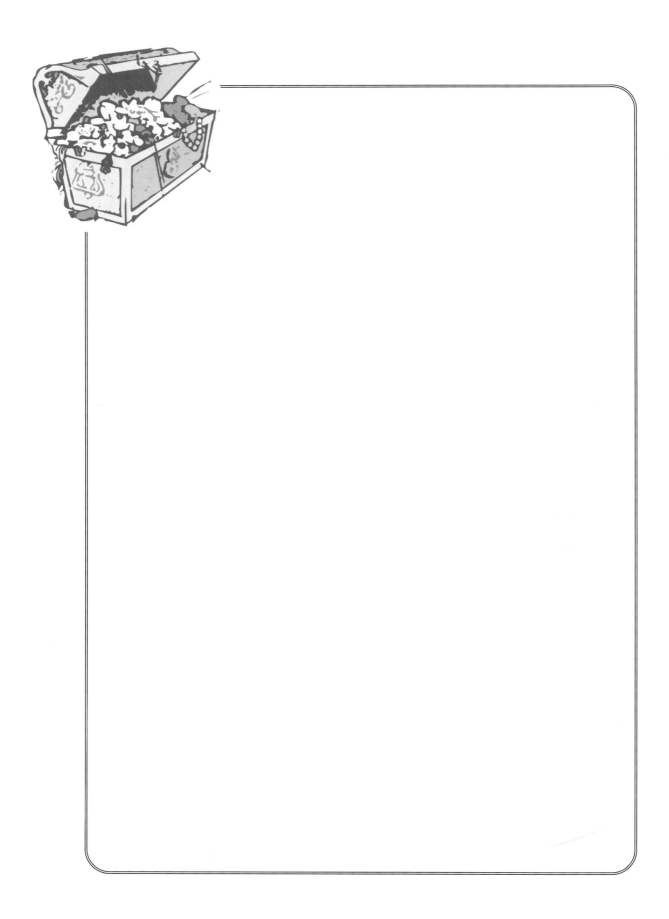

18.   Scallop

19.   Sea cucumber

20.   Sand dollar

21.   Sea anemone

22.   Octopus

23.   Squid

24.   Ray

25.   Manatee

VIII. Seashore   due – Sept 3

A.   Beaches

1.   Tidal pools

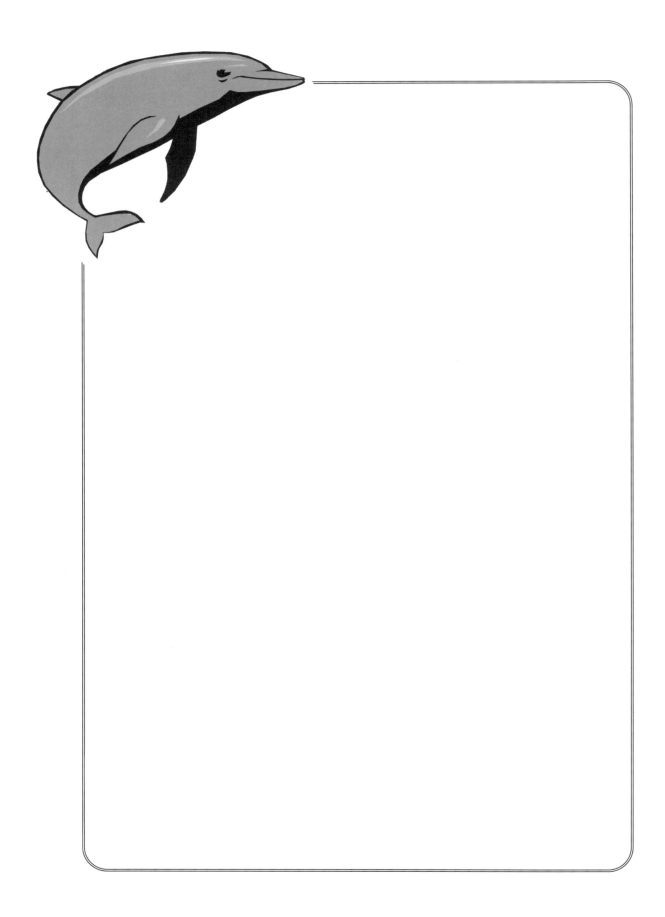

2. Sea shells

3. Sea birds

    a. Sea gull

    b. Pelican

    c. Sandpiper

    d. Heron

4. Land animals that live along the shore

    a. Sand crab (fiddler crab)

    b. Sand flea

5. Shoreline of the beach shifts through various tide changes

B. Recreation

1. Swimming

2.   Boating

3.   Surfing

4.   Fishing

5.   Scuba diving/snorkeling

6.   Shell collecting

7.   Treasure hunting

IX.   Weather and the ocean   *—due Sept 10*

A.   Hurricanes

B.   Tsunamis

C.   Typhoons

D.   Water spouts

E.   El Niño

F.   Prevailing winds

X.   Lighthouses   *due - sept 7*

A.   Purpose

B.   Types

C.   History

D.   Location

XI.   Art and ocean

A.   Literature

    1.   *Moby Dick*

    2.   *Swiss Family Robinson*

3.   *Robinson Crusoe*

4.   *Treasure Island*

5.   *Kidnapped*

6.   *20,000 Leagues Under the Sea*

B.   Paintings

    1.   Seascapes

    2.   Ships at sea

C.   Music

    1.   Sailing songs

    2.   Sounds of the ocean set to music

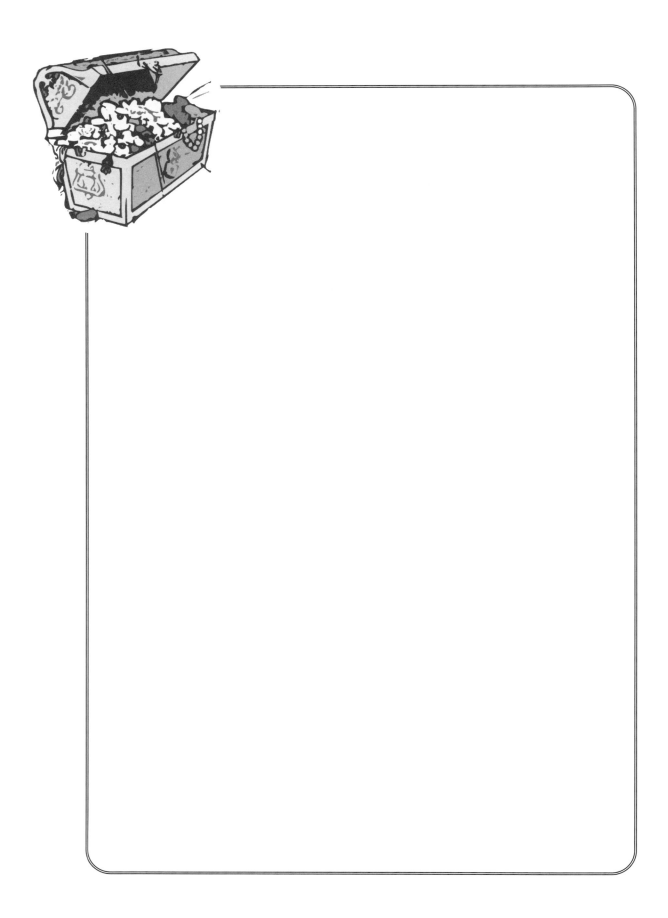

XII. Modern ocean research

A.   Goals

    1.   Resources

        a.   Food

        b.   Petroleum

        c.   Minerals

    2.   Information to improve medical technology

    3.   Increase our knowledge of the ocean floor and its components

B.   Research institutes continue the quest for knowledge of the ocean

    1.   Scripps Institution of Oceanography

    2.   Woods Hole Oceanographic Institution

    3.   Harbor Branch Oceanography Institute

About The Author

Amanda Bennett, author and speaker, wife and mother of three, holds a degree in mechanical engineering. She has written this ever-growing series of unit studies for her own children, to capture their enthusiasm and nurture their gifts and talents. The concept of a thematic approach to learning is a simple one. Amanda will share this simplification through her books, allowing others to use these unit study guides to discover the amazing world that God has created for us all.

Science can be a very intimidating subject to teach, and Amanda has written this series to include science with other important areas of curriculum that apply naturally to each topic. The guides allow more time to be spent enjoying the unit study, instead of spending weeks of research time to prepare for each unit. She has shared the results of her research in the guides, including plenty of resources for areas of the study, spelling and vocabulary lists, fiction and nonfiction titles, possible careers within the topic, writing ideas, activity suggestions, addresses of manufacturers, teams, and other helpful resources.

The science-based series of guides currently includes the Unit Study Adventures titles:

Baseball	Home
Computers	Oceans
Elections	Olympics
Electricity	Pioneers
Flight	Space
Gardens	Trains

The holiday-based series of guides currently includes the Unit Study Adventures titles:

Christmas
Thanksgiving

This planned 40-book series will soon include additional titles, which will be released periodically. We appreciate your interest. "Enjoy the Adventure."